TONY CLOSE

Microelectronics NIII

3

Gary McKenna.

NORTH EAST
INSTITUTE
OF FURTHER AND HIGHER
EDUCATION

MAGHERAFELT CAMPUS
22 Moneymore Road, Magherafelt,
Co. Londonderry, Northern Ireland, BT45 6AE.
Tel. (01648) 32462 Fax (01648) 33501

Also from Stanley Thornes

MICROELECTRONIC SYSTEMS – G.M. Cornell
MICROPROCESSOR INTERFACING – G. Dixey
MICROELECTRONICS NII – D. Turner
PRACTICAL EXERCISES IN MICROELECTRONICS – D. Turner

Microelectronics NIII

David Turner BSc(Eng) CEng MIEE ACGI

Dean of the Faculty of Technology
College of Further Education, Plymouth

Stanley Thornes (Publishers) Ltd

First published in 1991 by:
Stanley Thornes (Publishers) Ltd
Ellenborough House
Wellington Street
CHELTENHAM GL50 1YD
England

Reprinted 1993
Reprinted 1994

British Library Cataloguing in Publication Data
Turner, David
   Microelectronics NIII.—(Microelectronics)
   I. Title   II. Series
   621.381

   ISBN 0–7487–1177–5

Typeset by Florencetype Ltd, Kewstoke, Avon.
Printed and bound in Great Britain at The Bath Press, Avon.

Contents

Foreword

In 1982 Hutchinson Educational Publishers, now part of Stanley Thornes, published on behalf of the Business Technician Education Council (BTEC), a series of books designed for use as learning packages in association with the published standard units in Microelectronic Systems and Microprocessor-based Systems.

The last decade has seen the transformation of industrial computing with the explosion in personal computing. The reduction in price complemented by a significant increase in computing power has extended the application of personal computers so that microelectronics is now a realistic tool in all sectors of industry and commerce. The need for adequate training programmes for technicians and engineers has increased and the BTEC units have been revised and updated to reflect today's needs.

Stanley Thornes have produced a series of learning packages to support the updated syllabuses and numerous other courses which include Microelectronics and Microprocessor-based Systems. There are five books in the series:

Microelectronics Systems	Level F	by G. Cornell
Microelectronics NII	Level N	by D. Turner
Microelectronics NIII	Level N	by D. Turner
Microprocessor-Based Systems	Level H	by R. Seals
Microcomputer Systems	Level H	by R. Seals

Two additional books which complement the above five are:

Microprocessor Interfacing	Level N	by G. Dixey
Practical Exercises in Microelectronics	Level N	by D. Turner

This book follows the BTEC unit and builds on the material developed in the complementary *Microelectronics NII* to cover totally microelectronics and micro-processor fundamentals for N level BTEC students. Further books in the series by R. Seals extend the subject matter to more advanced system studies. *Practical Exercises in Microelectronics* complements this and the level NII book.

Andy Thomas
Series Editor

Preface

This book is written to cover the objectives of the BTEC Unit Microelectronic Systems U86/333 which are specified for a single unit at NIII level. The companion volume in the series covers the other objectives of the same unit at NII level. In addition the two books are supported by a third, entitled *Practical Exercises in Microelectronics* which provides a large number of relevant practical exercises designed specifically for student use. The three Microelectronics books cover all aspects of the BTEC Unit from a theoretical and practical point of view providing material for both classroom activities and self-supported study.

For those students who prefer to study by distance- or open-learning methods, two Learning Packages are available which integrate the theory and the appropriate practical activities. The Packages also contain all the required hardware and software needed for a complete microelectronics course. Details are available from:

Plymouth Open Learning Systems Unit
College of Further Education
Kings Road
Devonport
Plymouth
PL1 5QG

Telephone: (0752) 551947

The aims of this book are to extend the student's understanding of the operation of microprocessor-based systems by examining the data sheets and typical use of common integrated circuits found in microcomputers. It also aims to develop the techniques of assembly language programming introduced in the previous volume, with particular reference to the special requirements of interrupt-driven systems. Practical aspects of microcomputers are covered in detail, including the design considerations for microcomputer printed circuit boards.

Since this is not an introductory book, it assumes a certain basic knowledge. In particular, it assumes that the student has studied the objectives specified for the BTEC Microelectronic Systems Unit at NII level. These include basic microcomputer system and CPU architecture, the fetch-executive cycle, basic machine code programming and the use of subroutines. It also assumes that binary numbers, hexadecimal numbers and instruction mnemonics are fully understood. If there is any doubt about these topics it is recommended that the Microelectronics NII textbook be studied first as an introduction.

Throughout this book and the others in the series, the majority of the examples and all of the programs use the Z80 microprocessor. Not only is the Z80 the most widely-used 8-bit microprocessor, but it also illustrates most of the aspects of microprocessor operation which are in widespread use throughout industry. Its features, such as the large number of registers, many addressing modes, non-multiplexed buses and the range of interrupt facilities, permit detailed examination of these microprocessor features with reference to a single device. The large number of Z80-based computer systems in educational establishments can also be employed to run the programs contained in this and the other books in the series.

David Turner
September 1991

Acknowledgements

The author wishes to thank his wife Kathy, whose help and encouragement during the production of this book have been a constant source of inspiration and whose sense of humour has somehow managed to remain intact throughout. Thanks also go to his understanding children Beth, Alex and Ben, who should have seen a lot more of their father than they have done.

His thanks go to Jackie Boyce who typed and edited the manuscript so efficiently and whose typing speed is a good test for any wordprocessor. Thanks also go to Roger Bond and Elizabeth Frederick-Preece of POLSU who created some excellent illustrations from his very rough sketches.

ONE

Practical micro-electronic circuits

When you have completed this chapter, you should be able to:

1. *Appreciate the characteristics of the main integrated circuit technologies.*
2. *Explain the operation of address bus buffers and address decoders.*
3. *Relate the use of memory address decoders to the system memory map.*
4. *Appreciate the special requirements of dynamic memory devices.*

1.1 INTEGRATED CIRCUIT TECHNOLOGY

The development of the transistor in 1947 marks the beginning of the revolution in microelectronics which we know today. The transistor forms the basis of all the electronic circuits that are found in a digital computer, although several types of transistor now exist.

Integrated circuits were first produced by combining several transistors on a single piece of silicon. At first, only a few transistors were employed, and the integrated circuits were known as **SSI (small-scale integration)** devices. However, development was quite rapid in the number of transistors that were placed on the same piece of silicon, and by the late 1960s many thousands of transistors were being employed in single integrated circuits. In the relentless effort to pack more and more transistors on the same piece of silicon, inevitably the original **bipolar transistors** made way for devices known as **field effect transistors (FETs)**. These devices consumed less power than bipolar transistors, and could be designed in a smaller area of silicon, thus allowing much greater packing densities to be achieved.

As with transistors, which come in **PNP** and **NPN** types, the FETs come in P-channel and N-channel types. Generally these are known as **PMOS** and **NMOS** types, respectively. Here, **MOS** refers to Metal Oxide Silicon, which is the fabrication method for the field effect transistors used.

Circuits that are relatively simple such as logic gates require only a few transistors on a chip, and are known as **SSI (small scale-integration)** or **MSI (medium-scale integration)** devices. Typically an SSI device may have up to 10 transistors on a single integrated circuit. Similarly, **MSI** devices may have up to 100 transistors.

The most common MSI devices in use in modern computer systems are the 74 Series of logic devices. The most basic of these are based on normal bipolar transitors (**NPN**) but they have been developed using a type of transistor known as a **Schottky** device which has a higher operating speed and can be made to operate using little power. Advanced low power Schottky devices (**ALS**) also exist, which can operate even faster and have a lower power dissipation.

Transistor–transistor logic – TTL

Transistor–transistor logic (TTL) devices dominated the world of digital electronics for many years. They comprise a wide range of logic devices which cover many operating functions required within digital circuits, and a TTL device exists for most functions. They are therefore the building blocks of many digital systems. Most TTL devices come in 14, 16, 18, 20 or 24 pin packages, and all employ a standard power supply voltage of +5 volts. The devices have been developed over several years from the standard 74 series, to the low power Schottky series, and the advanced low power Schottky series of devices. *Table 1.1* gives a brief comparison of their relative merits.

TTL devices in each series are designed to have common input and output characteristics. This means that they employ similar input circuits throughout the series, no matter how complex the internal chip operation, as well as common output driving circuits. *Figure 1.1* shows a simple TTL inverter in both normal TTL and LS technologies. This device, particularly the LS type, is frequently used in microcomputer circuits.

Emitter coupled logic – ECL

Another family of logic devices that uses bipolar transistors is known as **emitter coupled logic**. This is because it is based on a differential amplifier circuit in which two transistors share a common emitter resistor. During operation the transistors do not saturate, unlike TTL, and therefore their switching speed can be made much faster.

The main characteristic of ECL is its speed, which makes it very popular with the designers of larger computers. It is not often used in micropro-cessor based systems. Typically the propagation delay of an ECL gate is 1 ns.

The major disadvantage of ECL as far as the microcomputer designer is concerned is the requirement for a supply voltage of -5.2 V. This also gives logic levels of -0.8 V (high) and -1.6 V (low) respectively. Clearly these are not TTL compatible and, because they are close together, ECL does not have very good noise immunity. This means that it cannot be used in electrically noisy environments, and the PCB design is critical. Typically, ECL gates dissipate 25–60 mW per gate and have a fan-out (load-driving figure) of 30. The basic ECL logic functions are based on the **OR/NOR** circuit shown in *Figure 1.2*.

Transistors T_1, T_2 and T_3 share a common emitter resistor R_3, and the base of T_3 is held at a fixed bias voltage. If the inputs A and B are both at a logic 0 level, both T_1 and T_2 are cut off and T_3 conducts. Thus T_2 collector is high and output x is also high, while output $\bar{x}$ is low. If either input A OR B go high causing T_1 or T_2 to conduct, the current through R_3 is diverted through the conducting transistor and T_3 turns off. The outputs therefore change state. Since T_1 OR T_2 can cause the same effect an **OR/NOR** function is generated at the outputs as shown.

CMOS

Soon after the development of TTL devices, a number of problems were apparent with the range of devices. For example, they were relatively sensitive in changes to supply voltage, they dissipated a relatively large power and although their rise and fall times were fast, the noise immunity was relatively poor. These and other considerations led to the development of an alternative logic family based on both P and N-channel field effect

Table 1.1 Comparison of TTL technologies

	Speed/power product	*Propagation delay*	*Power dissipation*	*Clock input frequency range*
74 TTL	100 pJ	10 ns	10 mW	DC – 35 MHz
74 LS TTL	19 pJ	9.5 ns	2 mW	DC – 45 MHz
74 ALS TTL	4 pJ	4 ns	1 mW	DC – 50 MHz

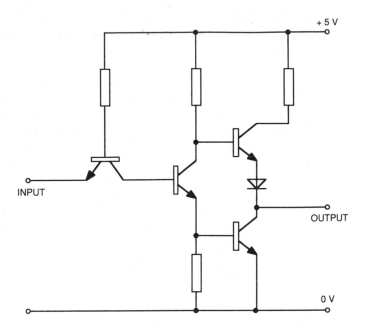

Figure 1.1(a)   Standard TTL inverter

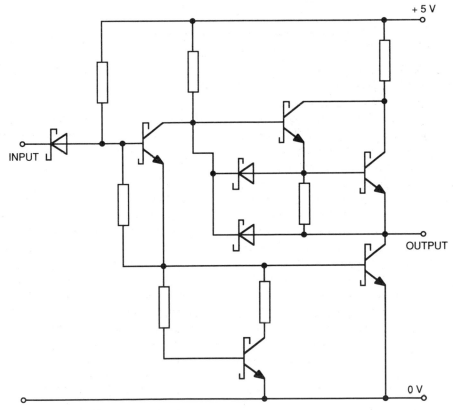

Figure 1.1(b)   LS TTL inverter

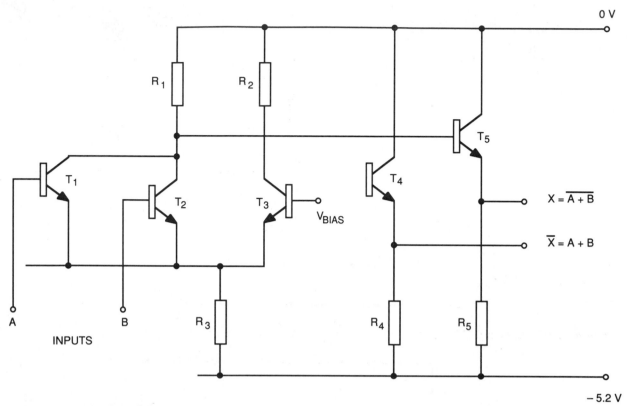

Figure 1.2 ECL OR/NOR gate

transistors. Together these are known as **Complementary MOS** devices or **CMOS** for short. CMOS logic families exist in much the same way as the TTL logic families and they are known as the 4000 series. In addition there are some pin-compatible devices with the 74 logic series which makes them interchangeable. Whichever pin connections are used, CMOS devices are very widely used throughout the computer industry. They dissipate very little power because of their design and are therefore popular in circuits that need to be powered by batteries or have very low power requirements for some reason. They are also relatively fast but not generally as fast as TTL devices. However they can be operated from a wide range of voltages, typically between +3 and +18 V and this makes their operation very flexible. Because they are based on field effect transistors, their input impedance is very large which means that their fanout capability is very large with one gate typically driving up to 50 others.

CMOS devices are relatively complex to manufacture, since both P and N-channel devices are incorporated in the same silicon slice. This means that they are slightly more expensive on a chip-for-chip basis than TTL devices but in many situations they have advantages that TTL devices cannot hope to match. Their only real disadvantage is in their speed of operation which is marginally slower than the comparable TTL devices but still fast enough for many applications.

Many devices have been produced using CMOS and it is very common to see CMOS microprocessor chips and CMOS memory chips which are specifically designed for battery powered equipment. However, these are usually relatively expensive.

Figure 1.3 shows the diagram of the internal structure of a typical CMOS gate. The device shown is a two-input **NAND** gate and this requires only four active devices.

In the NAND gate the N-channel devices are the

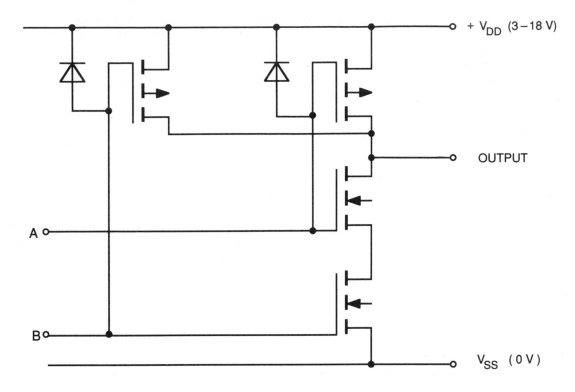

Figure 1.3 CMOS **NAND** gate

two in series between the output and the 0 volts supply. The P-channel devices are in parallel between the output and the V_{DD} supply. It is only when both inputs A and B are taken to a high logic level that the two N-channel devices in series turn on and therefore produce a low output. For all other logic conditions the output remains high.

Table 1.2 shows a typical comparison between the 74LS Series and the CMOS devices operating from the same supply voltage together with ECL.

NMOS

While it is perfectly possible to manufacture microprocessors and memory devices using CMOS technology, the quest for greater reliability, greater packing density and lower cost of manufacture has inevitably led to the production of circuits using only one type of FET. This is the N-channel MOS device using NMOS technology. The vast majority of microprocessors and

Table 1.2 LS TTL, CMOS and ECL comparisons

	Speed/power product	Propagation delay	Power dissipation	Clock input frequency range
74 LS TTL	19 pJ	9.5 ns	2 mW	DC – 45 MHz
CMOS (5 V)	5×10^{-4} pJ	50 ns	10^{-5} mW	DC – 5 MHz
ECL	50 pJ	1 ns	50 mW	DC – 150 MHz

memory chips are now manufactured using NMOS devices. The fact that semiconductor memories can be manufactured at lower cost and in high volume has made the cost of computing less expensive at all levels. Low-cost memory has reduced the cost of mainframes and mini-computers, and low-cost memory devices have been the keystone in the development of micro-computers. It is the flexibility of NMOS that has made this type of development possible in **dynamic RAMs**, **static RAMs** and **ROMs**.

Modern NMOS devices operate from a single +5 V supply, the same as TTL devices, and can thus be interfaced relatively easily. In addition their operating speed is comparable with TTL and this makes them very attractive for the logic circuit designer.

Figure 1.4 shows some typical applications of NMOS devices which may be part of a memory device or a microprocessor. Notice in *Figure 1.4*, that the symbols used for the NMOS devices are much simpler than those for NMOS devices in *Figure 1.3*. This is because when only NMOS devices are being considered, it is far easier to draw a simpler diagram for the device.

Many modern microprocessor systems will contain a mixture of TTL, CMOS, and NMOS devices, with the TTL and NMOS in greatest abundance. For example, a typical microcomputer will contain a NMOS microprocessor with NMOS memory devices, but the other peripheral chips within the system such as the decoders and bus drivers are likely to be TTL circuits. Frequently CMOS memory chips are used for 'battery backed RAM'. This is a special part of the memory which is used to store operating system parameters such as the number of disk drives and the type of memory installed, and it is important that this information is not altered when the power is removed. However, it cannot be held in ROM because the information must be changeable.

1.2 THE OPERATION OF THE ADDRESS BUS

One feature that all computers have in common is that they are equipped with memory devices that are used to store both programs and data. In addition, all computers need some form of input/output device to allow them to communicate with other systems. The tendency with most modern computers is to have a very large memory capacity so that very complex programs can be stored and executed.

Each location in a large memory system and each part of an input/output device must be capable of being uniquely specified so that its data or operation may be manipulated by the CPU. Therefore, each part of the system must have a unique **address**.

Addresses are specified in binary as a number placed on the CPU address bus. Since the number of unique addresses is related to the number of address lines with the formula:

$$\text{Addresses} = 2^{\text{No. of address lines}}$$

clearly the larger the number of address lines the larger the number of addresses that are allowed in a system:

16 address lines =   65536 addresses
20 address lines =   1048576 addresses
24 address lines =   16777216 addresses

Fortunately, most 8 bit microprocessors have only 16 address lines and therefore the maximum memory that they support is 64K (65 536 bytes). This makes the process of understanding how the address bus works relatively easy. However, exactly the same principles apply even when considering 16-bit or 32-bit CPUs with much wider address buses.

In simple systems, the CPU is the only device that can place an address on the address bus. Therefore it is said to be **unidirectional** since all addresses are output by the CPU and input by the memory and input/output devices. Some larger systems include devices such as **direct memory access controllers** which are also capable of placing addresses on the address bus. However, in these cases the CPU relinquishes control of the bus while another device is controlling it, to prevent any conflict arising. This is the reason that the address bus outputs must be capable of **tri-state** operation under certain circumstances.

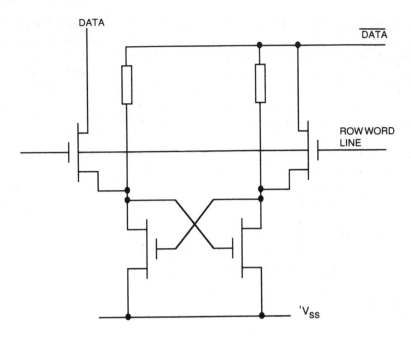

Figure 1.4(a) Part of an NMOS static RAM

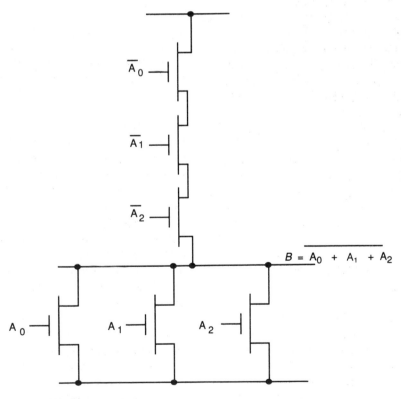

$$B = \overline{A_0 + A_1 + A_2}$$

Figure 1.4(b) NMOS logic function

Address bus loading

Memory devices are manufactured in many different sizes and configurations. These include $2K \times 8$ bits, $4K \times 8$ bits, $16K \times 8$ bits, $16K \times 1$ bit, $64K \times 1$ bit, $256K \times 1$ bit, etc.

Part of the work of the system designer is to choose suitable memory devices to provide a system with sufficient memory of the right type which operates at the right speed.

This means that in some cases a large number of memory devices may be connected to the address bus. If each of these devices acts as a load on the CPU address bus pins then with only a few devices connected, the CPU output logic levels could be affected. For example, the output drive capability of a Z80 CPU is given in the data book as:

Logic 1 – 250 μA (source)
Logic 0 – 1.8 mA (sink)

This means that it has the drive capability for 1 normal TTL load or 4 LS TTL loads.

While this drive capability may be more than adequate for small systems, in larger systems with many memory and input/output devices, some means of increasing this drive capability must be provided. This will ensure that all the logic levels are maintained at their design levels even at the highest operating speed of the system.

The usual method of increasing the drive capability of the CPU is to include an extra device known as a **bus buffer** or **bus driver** between the CPU address bus pins and the rest of the address bus connections. This is shown in *Figure 1.5* (opposite). If OCTAL devices are used then two could be required for the 16 address lines.

1.3 TYPICAL ADDRESS BUS BUFFERS

Many systems employ address bus buffers, or line drivers from the 74 series of logic devices. This contains a range of chips suitable for most applications including both inverting and non-inverting outputs and active high or active low enable inputs.

Three of these devices that find frequent application both as bus drivers and input/output ports

are the 74LS240, 74LS241 and the 74LS244. *Figure 1.6* shows the data sheets for these devices. Notice also that some data is given for the 54 series of logic devices which are designed to military specification, with a much wider temperature range. Each device is a 20-pin DIL package with two groups of four drivers each controlled by a separate control input..

A number of important and interesting facts can be obtained from these data sheets which help to explain the operation of the circuits.

The **drive capability** is the first point of interest. Figures are given for output currents I_{OL} and I_{OH} on the first page.

For a 74LS device:

$$I_{OL} = 24 \text{ mA}$$

and $$I_{OH} = -15 \text{ mA}$$

This compares with the figures given earlier for the Z80 of:

$$I_{OL} = 1.8 \text{ mA}$$

and $$I_{OH} = -250 \text{ μA}.$$

The **bus driver** is therefore capable of providing a significantly higher current at both logic levels and as such has a very large fan-out capability.

The **noise margin** for the gates is increased because of the hysteresis designed into the circuits. The greater the **noise margin** of a gate, the more noisy the signal can be before the output changes state. In the case of this series of chips, the noise margin is 400 mV.

The **propagation delay** through the gate varies between 9 ns and 18 ns depending upon the device and whether it is a rising or falling edge under consideration. This propagation delay must be added to any other delays in the address bus chips when considering how quickly the memory devices can be expected to provide data. For example, with a 4 MHz CPU clock, each clock cycle is only 250 ns, or 125 ns per half cycle, so the delay of 18 ns could become significant if other delays are also taken into consideration.

The **input current** requirement is only 20 μA at logic 1 and -0.2 mA at logic 0, so the Z80 has plenty of drive available for the buffer inputs.

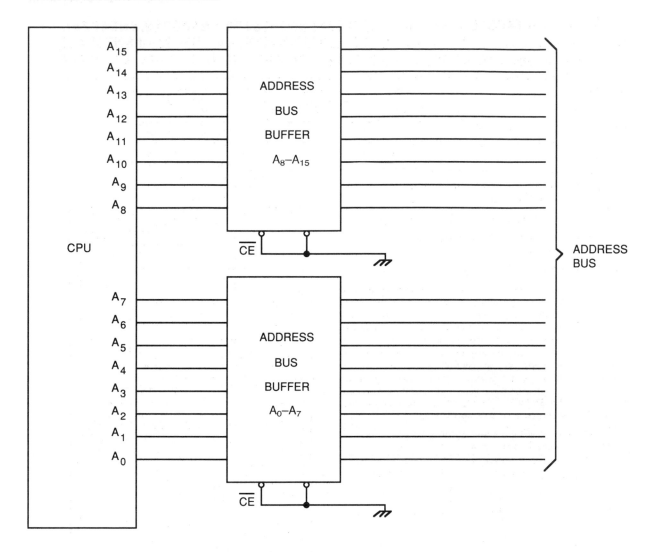

Figure 1.5 Use of address bus buffers

1.4 ADDRESS DECODING

Most computer memories contain a large number of devices, and these are interconnected to form the whole memory array. Normally most of these devices are of the same type although it is possible to have a range of different chips in one system.

To create a complete memory map, each chip must be allocated a range of memory addresses which are unique. This means that only one chip will respond with data when a particular address is placed on the address bus.

The process that ensures this unique response to a given address is known as **address decoding** and it may be carried out in a variety of ways. Each method uses simple logic circuits which select only one memory device for each address.

EXAMPLE 1

Consider the problem of how to connect two 1K × 8 bit memory chips into a system so that they will have consecutive ranges starting at 0000 hex and extending to 07FF hex (*Figure 1.7*). Each device contains 1024 memory locations and the range of

TYPES SN54LS240, SN54LS241, SN54LS244, SN54S240, SN54S241, SN74LS240, SN74LS241, SN74LS244, SN74S240, SN74S241
OCTAL BUFFERS AND LINE DRIVERS WITH 3-STATE OUTPUTS

	Typical I_{OL} (Sink Current)	Typical I_{OH} (Source Current)	Typical Propagation Delay Times		Typical Enable/ Disable Times	Typical Power Dissipation (Enabled)	
			Inverting	Noninverting		Inverting	Noninverting
SN54LS'	12 mA	−12 mA	10.5 ns	12 ns	18 ns	130 mW	135 mW
SN74LS'	24 mA	−15 mA	10.5 ns	12 ns	18 ns	130 mW	135 mW
SN54S'	48 mA	−12 mA	4.5 ns	6 ns	9 ns	450 mW	538 mW
SN74S'	64 mA	−15 mA	4.5 ns	6 ns	9 ns	450 mW	538 mW

- **3-State Outputs Drive Bus Lines or Buffer Memory Address Registers**
- **P-N-P Inputs Reduce D-C Loading**
- **Hysteresis at Inputs Improves Noise Margins**

description

These octal buffers and line drivers are designed specifically to improve both the performance and density of three-state memory address drivers, clock drivers, and bus-oriented receivers and transmitters. The designer has a choice of selected combinations of inverting and noninverting outputs, symmetrical $\overline{G}$ (active-low output control) inputs, and complementary G and $\overline{G}$ inputs. These devices feature high fan-out, improved fan-in, and 400-mV noise-margin. The SN74LS' and SN74S' can be used to drive terminated lines down to 133 ohms.

schematics of inputs and outputs
'LS240, 'LS241, 'LS244

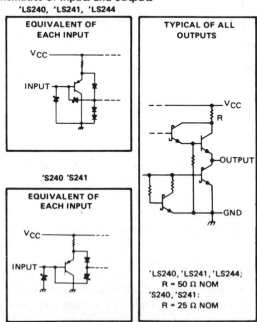

'S240 'S241

'LS240, 'LS241, 'LS244;
R = 50 Ω NOM
'S240, 'S241:
R = 25 Ω NOM

SN54LS240, SN54S240 . . . J
SN74LS240, SN74S240 . . . J OR N
(TOP VIEW)

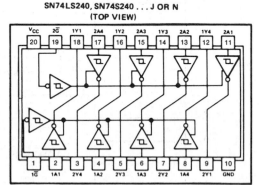

SN54LS241, SN54S241 . . . J
SN74LS241, SN74S241 . . . J OR N
(TOP VIEW)

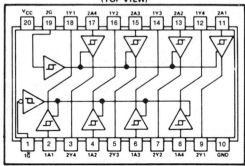

SN54LS244 . . . J
SN74LS244 . . . J OR N
(TOP VIEW)

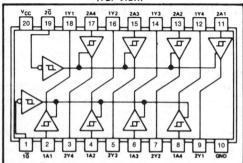

Figure 1.6(a)–(c)   Bus driver data sheets (Courtesy Texas Instruments)

TYPES SN54LS240,SN54LS241,SN54LS244, SN74LS240,SN74LS241,SN74LS244 BUFFERS AND LINE DRIVERS WITH 3-STATE OUTPUTS

recommended operating conditions

PARAMETER	SN54LS' MIN	SN54LS' NOM	SN54LS' MAX	SN74LS' MIN	SN74LS' NOM	SN74LS' MAX	UNIT
Supply voltage, V_{CC} (see Note 1)	4.5	5	5.5	4.75	5	5.25	V
High-level output current, I_{OH}			−12			−15	mA
Low-level output current, I_{OL}			12			24	mA
Operating free-air temperature, T_A	−55		125	0		70	°C

NOTE 1: Voltage values are with respect to network ground terminal.

electrical characteristics over recommended operating free-air temperature range (unless otherwise noted)

PARAMETER		TEST CONDITIONS[†]		SN54LS' MIN	SN54LS' TYP[‡]	SN54LS' MAX	SN74LS' MIN	SN74LS' TYP[‡]	SN74LS' MAX	UNIT
V_{IH}	High-level input voltage			2			2			V
V_{IL}	Low-level input voltage					0.7			0.8	V
V_{IK}	Input clamp voltage	V_{CC} = MIN,	I_I = −18 mA			−1.5			−1.5	V
	Hysteresis ($V_{T+} - V_{T-}$)	V_{CC} = MIN		0.2	0.4		0.2	0.4		V
V_{OH}	High-level output voltage	V_{CC} = MIN, V_{IL} = V_{IL} max,	V_{IH} = 2 V, I_{OH} = −3 mA	2.4	3.4		2.4	3.4		V
		V_{CC} = MIN, V_{IL} = 0.5 V,	V_{IH} = 2 V, I_{OH} = MAX	2			2			
V_{OL}	Low-level output voltage	V_{CC} = MIN, V_{IH} = 2 V, V_{IL} = V_{IL} max	I_{OL} = 12 mA			0.4			0.4	V
			I_{OL} = 24 mA						0.5	
I_{OZH}	Off-state output current, high-level voltage applied	V_{CC} = MAX, V_{IH} = 2 V,	V_O = 2.7 V			20			20	μA
I_{OZL}	Off-state output current, low-level voltage applied	V_{IL} = V_{IL} max,	V_O = 0.4 V			−20			−20	
I_I	Input current at maximum input voltage	V_{CC} = MAX,	V_I = 7 V			0.1			0.1	mA
I_{IH}	High-level input current, any input	V_{CC} = MAX,	V_I = 2.7 V			20			20	μA
I_{IL}	Low-level input current	V_{CC} = MAX,	V_{IL} = 0.4 V			−0.2			−0.2	mA
I_{OS}	Short-circuit output current[♦]	V_{CC} = MAX		−40		−225	−40		−225	mA
I_{CC}	Supply current	Outputs high, V_{CC} = MAX	All		17	27		17	27	mA
		Outputs low	'LS240		26	44		26	44	
		Outputs open	'LS241, 'LS244		27	46		27	46	
		All outputs disabled	'LS240		29	50		29	50	
			'LS241, 'LS244		32	54		32	54	

[†]For conditions shown as MIN or MAX, use the appropriate value specified under recommended operating conditions.
[‡]All typical values are at V_{CC} = 5 V, T_A = 25°C.
[♦]Not more than one output should be shorted at a time, and duration of the short-circuit should not exceed one second.

switching characteristics, V_{CC} = 5 V, T_A = 25°C

PARAMETER		TEST CONDITIONS		'LS240 MIN	'LS240 TYP	'LS240 MAX	'LS241, 'LS244 MIN	'LS241, 'LS244 TYP	'LS241, 'LS244 MAX	UNIT
t_{PLH}	Propagation delay time, low-to-high-level output	C_L = 45 pF, See Note 2	R_L = 667 Ω,		9	14		12	18	ns
t_{PHL}	Propagation delay time, high-to-low-level output				12	18		12	18	ns
t_{PZL}	Output enable time to low level				20	30		20	30	ns
t_{PZH}	Output enable time to high level				15	23		15	23	ns
t_{PLZ}	Output disable time from low level	C_L = 5 pF, See Note 2	R_L = 667 Ω,		15	25		15	25	ns
t_{PHZ}	Output disable time from high level				10	18		10	18	ns

NOTE 2: Load circuit and voltage waveforms are shown on page 3-11.

Figure 1.6(b)

TYPES SN54LS240,SN54LS241, SN54LS244,SN54S240,SN54S241,SN74LS240, SN74LS241,SN74LS244,SN74S240,SN74S241
OCTAL BUFFERS AND LINE DRIVERS WITH 3-STATE OUTPUTS

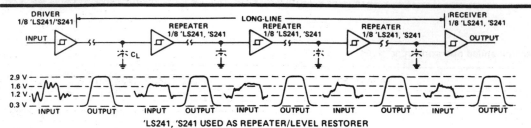

'LS241, 'S241 USED AS REPEATER/LEVEL RESTORER

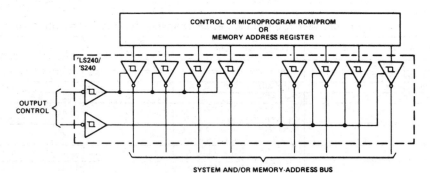

'LS241, 'S240 USED AS SYSTEM AND/OR MEMORY BUS DRIVER—4-BIT
ORGANIZATION CAN BE APPLIED TO HANDLE BINARY OR BCD

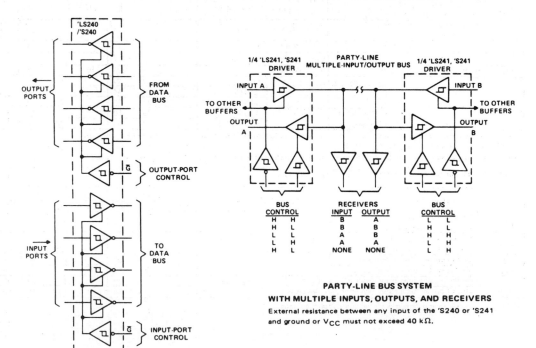

INDEPENDENT 4-BIT BUS DRIVERS/RECEIVERS
IN A SINGLE PACKAGE

PARTY-LINE BUS SYSTEM
WITH MULTIPLE INPUTS, OUTPUTS, AND RECEIVERS

External resistance between any input of the 'S240 or 'S241
and ground or V_{CC} must not exceed 40 kΩ.

Figure 1.6(c)

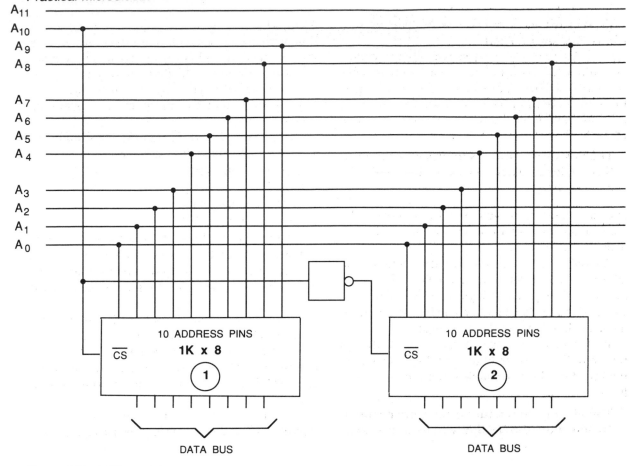

Figure 1.7 A 2K memory system

Hex address	15	14	13	12	11	10	9	8	7	6	5	4	3	2	1	0
0000	0	0	0	0	0	0	0	0	0	0	0	0	0	0	0	0
0001	0	0	0	0	0	0	0	0	0	0	0	0	0	0	0	1
03FE	0	0	0	0	0	0	1	1	1	1	1	1	1	1	1	0
03FF	0	0	0	0	0	0	1	1	1	1	1	1	1	1	1	1
0400	0	0	0	0	0	1	0	0	0	0	0	0	0	0	0	0
0401	0	0	0	0	0	1	0	0	0	0	0	0	0	0	0	1
07FE	0	0	0	0	0	1	1	1	1	1	1	1	1	1	1	0
07FF	0	0	0	0	0	1	1	1	1	1	1	1	1	1	1	1

Figure 1.8 2K memory addressing

addresses within each one can be expressed in binary as shown in *Figure 1.8*.

Each device has 10 address pins and these must be connected to the low-order address lines (A_0–A_9). From the chart it can be seen that exactly the same pattern of binary data appears on these pins throughout the address range 0000–03FF and 0400–07FF.

The low-order address lines select the location within each device that will be addressed.

Internal decoders in each memory chip select the appropriate memory location that will provide the data for the data bus.

The only change in the binary data between the address ranges 0000–03FF and 0400–07FF is that the second range has a logic 1 in the address bit 10 position. It is this fact that can be used to make the selection between the two chips.

In this case, a single inverter is all that is required. When A_{10} is **high** chip 2 is selected via the inverter. This is a simple **decoder**.

In general:

The higher order address lines are decoded to select the device that will be addressed.

EXAMPLE 2

Computer memories with more than two devices clearly need a rather more complex memory decoding arrangement than a single inverter. Suppose four 4K × 8 ROMs are used in a system to hold a large monitor program and operating system.

Each ROM will have 12 address lines, and they will occupy the following addresses:

ROM 1 – 0000 – 0FFF
ROM 2 – 1000 – 1FFF
ROM 3 – 2000 – 2FFF
ROM 4 – 3000 – 3FFF

It is clear that the lower 12 address lines will be connected in parallel between all the devices, and the high-order bits will need to be used to differentiate between the ROMs. This is shown in *Figure 1.9*.

It is only bits 12 and 13 of the address bus that change between the address ranges for each of the ROMs. Therefore a decoder which has **four** outputs would be required to select uniquely each chip. The **inputs** would have to be address lines A_{12} and A_{13}.

This requires a 2–4 line binary to decimal **decoder**. A suitable circuit is shown in *Figure 1.10*.

ROM	Hex address	15	14	13	12	11	10	9	8	7	6	5	4	3	2	1	0
0	0000	0	0	0	0	0	0	0	0	0	0	0	0	0	0	0	0
	0FFF	0	0	0	0	1	1	1	1	1	1	1	1	1	1	1	1
1	1000	0	0	0	1	0	0	0	0	0	0	0	0	0	0	0	0
	1FFF	0	0	0	1	1	1	1	1	1	1	1	1	1	1	1	1
2	2000	0	0	1	0	0	0	0	0	0	0	0	0	0	0	0	0
	2FFF	0	0	1	0	1	1	1	1	1	1	1	1	1	1	1	1
3	3000	0	0	1	1	0	0	0	0	0	0	0	0	0	0	0	0
	3FFF	0	0	1	1	1	1	1	1	1	1	1	1	1	1	1	1

Figure 1.9   Addressing four 4K × 8 ROMs

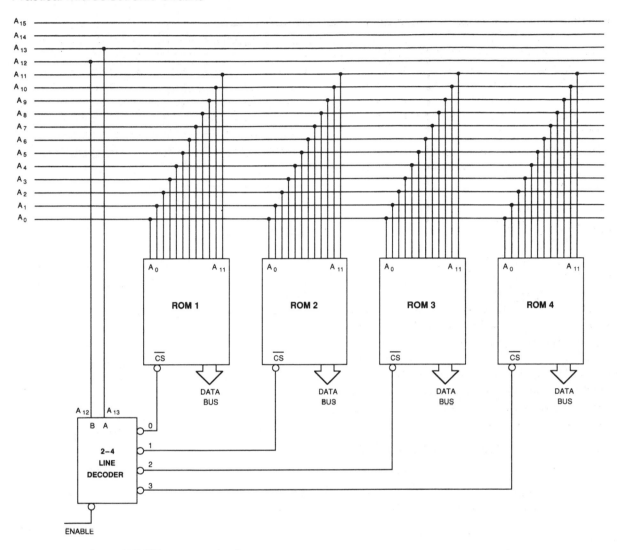

Figure 1.10   Four 4K ROM memory circuit

The **binary** code on the inputs of the **decoder** uniquely selects **one** of the outputs to be **low**. Most decoders also incorporate some type of **enable** input which must also be **low** to allow the selected output to go **low**.

This is shown in the **truth table** for such a decoder (*Figure 1.11*). Internally, decoders are simple combinational logic circuits which provide very high speed operating characteristics.

Figure 1.12 (overleaf) shows the internal logic of the simple 2–4 line decoder which could be used in the previous example. The same logic may

ENABLE	INPUTS		OUTPUTS			
	A	B	0	1	2	3
1	X	X	1	1	1	1
0	0	0	0	1	1	1
0	0	1	1	0	1	1
0	1	0	1	1	0	1
0	1	1	1	1	1	0

Figure 1.11   Decoder **truth table**

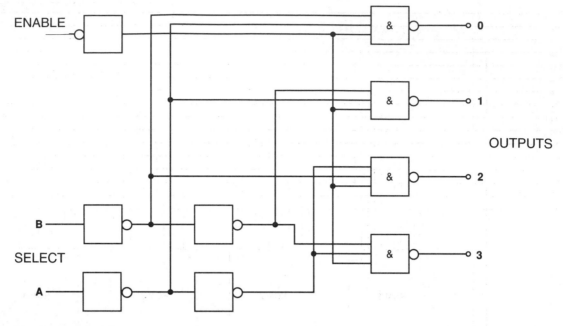

Figure 1.12 2–4 line decoder

be extended to provide 3–8-line and 4–16-line decoders, both of which are available as commercial chips.

Full and partial address decoding

In *Figure 1.9* and the circuit diagram in *Figure 1.10* it is clear that address bits A_{14} and A_{15} are not used. Therefore they could have any logic level and this would not affect the circuit operation.

For example, if the address 8000 hex was placed on the address bus, this would have exactly the same result as placing address 0000 hex on the bus. Bit 15 is ignored by the circuit.

A more careful examination of that circuit reveals that there are 4 groups of addresses which are all 'mirrors' of the base addresses 0000–3FFF. This is shown in the complete **memory map** (*Figure 1.13*).

When **partial** memory decoding is employed, the **real** memory appears to be repeated at other addresses. In this example, if data was written to address 0000 hex it could be read from addresses 4000 hex, 8000 hex or C000 hex.

Sometimes it may be useful for system designers to use this technique of **partial** decoding since in small systems it saves money and

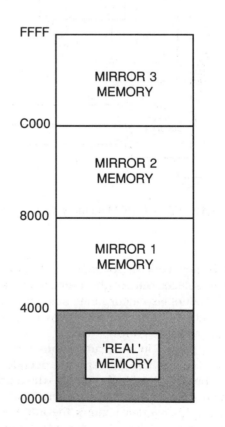

Figure 1.13 System memory map

space if all the decoders are not needed. However, in larger systems it is important that every memory address refers to a unique physically separate memory location.

Full address decoding utilises **all** of the address bus lines in the decoding process. For example, in *Figure 1.10*, if address lines A_{14} and A_{15} were combined in an OR gate whose output was connected to the *decoder enable* input, then the system would be **fully decoded**. This is because both of these lines would have to be at logic 0 before the decoder would work. If either of them went to a logic 1 then the system would not select any of the memory devices.

EXAMPLE 3

Most systems contain memory devices of different sizes. How could the decoding operation take place if a system were to contain four $2K \times 8$ RAM chips at consecutive addresses starting at 0000 hex, and four $8K \times 8$ ROM chips starting at address 2000 hex? Addresses should be fully decoded.

In this case, the starting point is to draw a **memory map** to obtain a clear picture of where the devices will appear (*Figure 1.14*). The range of addresses for each device can be seen from the map. A change in the **most significant hex** digit of 1 represents a change of 4K in the memory address.

The required decoding arrangement can be seen if the **binary** addresses are written down for each device as shown in *Figure 1.15* (overleaf). The RAM chips, each $2K \times 8$ require 11 address lines for the internal decoding. Therefore A_{11} and A_{12} may be used to select between them with a suitable decoder. However, this decoder must only operate when A_{13}, A_{14} and A_{15} are all logic 0. The ROM chips which are each $8K \times 8$ may be selected by decoding address lines A_{13}, A_{14} and A_{15}. Address lines A_0–A_{12} are used for internal decoding.

Therefore, **two** decoders are required: one 2–4 line and one 3–8 line. A suitable circuit is shown in *Figure 1.16* (overleaf).

Since all the memory devices are 8 bits wide, the 8 data lines are connected in parallel across all the chips.

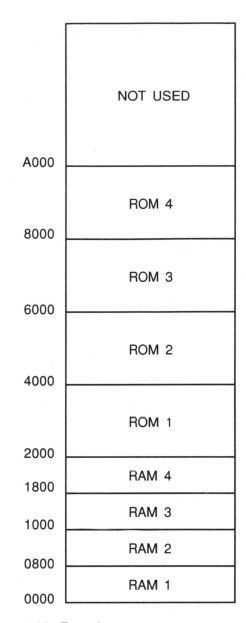

Figure 1.14   Example memory map

Linear address decoding

In small systems with very few memory devices it is sometimes possible to employ a technique known as **linear address decoding** which eliminates the need for the address decoders. It can save some of the circuit cost, but at the expense of a slightly more complex programming arrangement. Generally, it is only of value in ROM-based systems.

Device	Hex Address	15	14	13	12	11	10	9	8	7	6	5	4	3	2	1	0
RAM 1	0000	0	0	0	0	0	0	0	0	0	0	0	0	0	0	0	0
RAM 2	0800	0	0	0	0	1	0	0	0	0	0	0	0	0	0	0	0
RAM 3	1000	0	0	0	1	0	0	0	0	0	0	0	0	0	0	0	0
RAM 4	1800	0	0	0	1	1	0	0	0	0	0	0	0	0	0	0	0
ROM 1	2000	0	0	1	0	0	0	0	0	0	0	0	0	0	0	0	0
ROM 2	4000	0	1	0	0	0	0	0	0	0	0	0	0	0	0	0	0
ROM 3	6000	0	1	1	0	0	0	0	0	0	0	0	0	0	0	0	0
ROM 4	8000	1	0	0	0	0	0	0	0	0	0	0	0	0	0	0	0

Figure 1.15   Binary address ranges

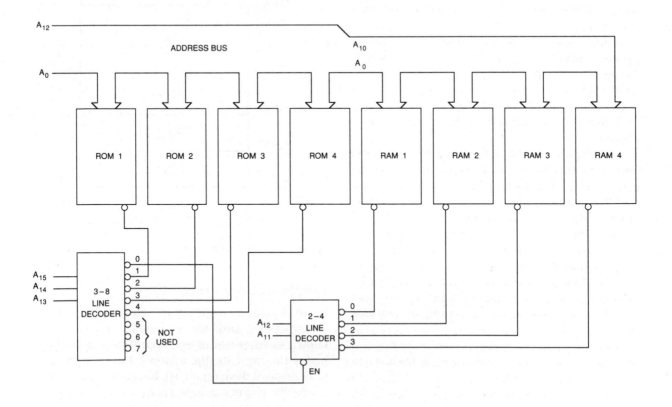

Figure 1.16   Example of decoded memory

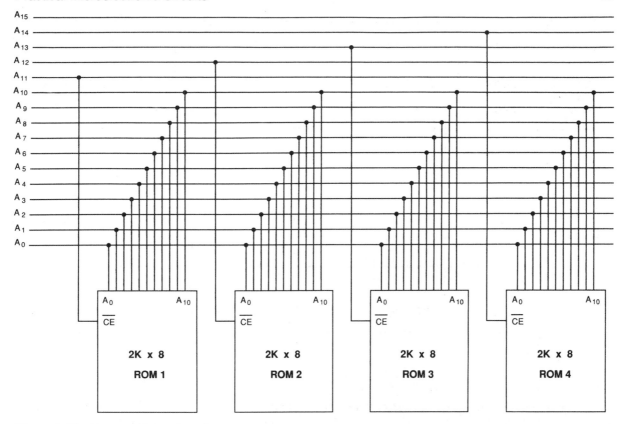

Figure 1.17   Linear address decoding

Instead of using the high-order address lines as inputs to an address decoder, they are used as the chip enable signals of the memory devices directly. This is shown in *Figure 1.17* where there are four $2K \times 8$ ROMs in the circuit. The $\overline{CE}$ signals for each device are derived from address lines A_{11}, A_{12}, A_{13} and A_{14}.

If these address lines are all at 0 simultaneously, all the chips would be addressed, and therefore this must be avoided. To cope with a system **reset** situation it is possible to start each device with the same **jump** instruction so that conflict is avoided.

To see how such a system could work, it is worthwhile working out the memory map from the binary addresses. It can be seen in *Figure 1.18* (overleaf) that the addresses are not contiguous, but that may not be important in some systems, as long as the software can be written that will accommodate this. Also, if one or more of the high-order address lines goes to 0 simultaneously then the system would probably fail catastrophically.

Generally, it is far better to employ some form of **address decoder**.

1.5 MEMORY-MAPPED INPUT/OUTPUT

The Z80 is one of a group of microprocessors that have separate signals to activate **memory** devices and **input/output** devices. Normally, the **memory** is selected by the **address decoder** and **enabled** by the $\overline{MREQ}$ signal. Frequently the $\overline{MREQ}$ signal is used as the chip enable signal for the address decoder chip. This ensures that the address decoder is only active when the memory request signal is active.

However, all the discussions so far about the addressing of memory devices could equally well apply to input/output (I/O) devices or **ports**, except that in most systems there are far fewer of them.

For example, a system with four parallel I/O

ROM	Hex address	15	14	13	12	11	10	9	8	7	6	5	4	3	2	1	0
1	7000	0	1	1	1	0	0	0	0	0	0	0	0	0	0	0	0
	77FF	0	1	1	1	0	1	1	1	1	1	1	1	1	1	1	1
2	6800	0	1	1	0	1	0	0	0	0	0	0	0	0	0	0	0
	6FFF	0	1	1	0	1	1	1	1	1	1	1	1	1	1	1	1
3	5800	0	1	0	1	1	0	0	0	0	0	0	0	0	0	0	0
	5FFF	0	1	0	1	1	1	1	1	1	1	1	1	1	1	1	1
4	3800	0	0	1	1	1	0	0	0	0	0	0	0	0	0	0	0
	3FFF	0	0	1	1	1	1	1	1	1	1	1	1	1	1	1	1

Figure 1.18　Linear decoding

devices may employ a decoder to select between them. *Figure 1.19* shows the $\overline{\text{IORQ}}$ signal activating the decoder. What would happen if the $\overline{\text{MREQ}}$ line were used, even though a port was being addressed?

The port would simply appear to be another memory address. When it was **read**, data would be **input**, and when it was **written** to, data would be **output**, assuming that it had been initialised correctly.

This technique is known as **memory mapped input/output**. Naturally the same addresses are not available for use as true memory, but that may be a very small price to pay for a simpler system. All the commands that work with memory addresses normally work with the I/O device. In fact some microprocessors do not have any separate **input/output** instructions in their instruction set, they simply rely upon the system using memory mapped I/O. Often, a block of **high-order** addresses are selected for I/O operations, as shown in *Figure 1.20* (page 22). The I/O decoder is only active when both A_{14} and A_{15} are at logic 1.

Notice that because of the fact the memory mapped addresses for I/O devices may not be fully decoded, it is likely that each port in such a system may occupy a range of possible addresses.

1.6 COMMERCIAL ADDRESS DECODERS

The TTL Series of logic devices has a number of **binary** to **decimal** decoders, in various sizes, including 2–4 lines, 3–8 line and 4–16 line. They also offer the possibility of normal TTL or open collector outputs, and both active high and active low configurations.

Commercial decoders may also be known as **demultiplexers**. The difference lies in the interpretation of the function of the pins of the device rather than its internal logic. In a demultiplexer its function is normally interpreted as 'routing the **enable** input state to one of the outputs, selected by the **binary** input code'. As a decoder, the device 'activates the output pin selected by the **binary** input code when correctly **enabled**'.

Typical devices includes the 74LS137, 74LS138, 74LS139, 74LS154, 74LS155, 74LS156, 74LS159, etc.

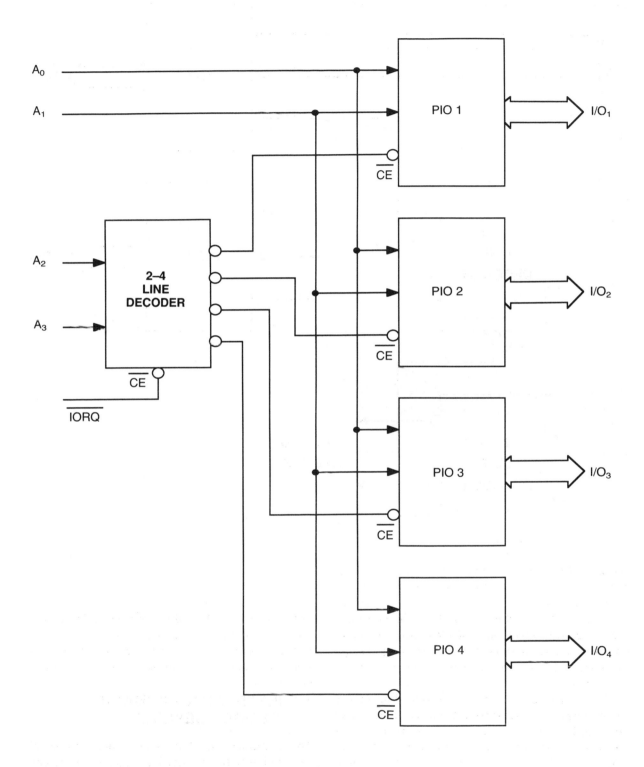

Figure 1.19 I/O selection with a decoder

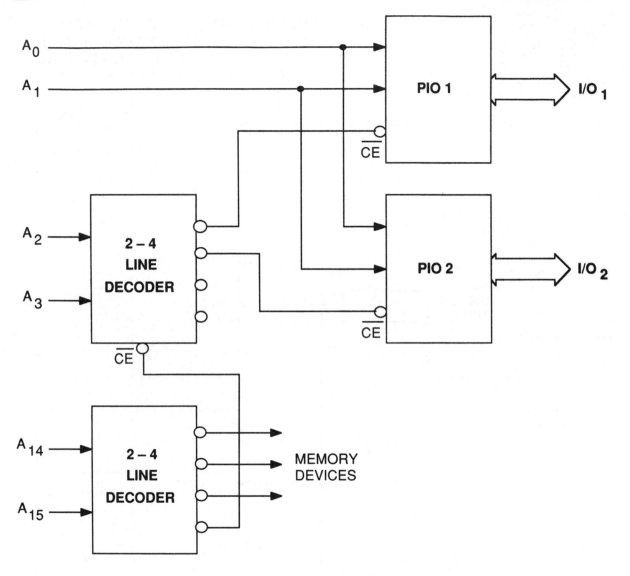

Figure 1.20 Memory mapped input/output

Probably the most widely used 2–4 line decoder is the 74LS139 which has a single enable input and comes as a **dual decoder** in a 16-pin package. The most frequently used 3–8 line decoder is probably the 74LS138.

The data sheets for both of these devices are given in *Figure 1.21*. Notice from the data sheets that both of these devices have 'active low' outputs, and that they have normal TTL totem pole output stages. The 74LS138 offers three enable inputs, of which two are active low and one is active high. In certain circumstances this can simplify the decoding arrangements.

The application of these devices can be seen in many commercial computer circuit diagrams.

1.7 INTERFACING DYNAMIC MEMORY DEVICES

The amount of Random Access Memory associated with microcomputer systems is ever increasing, and naturally more memory capacity is

TTL MSI

TYPES SN54LS138, SN54LS139A, SN54S138, SN54S139, SN74LS138, SN74LS139A, SN74S138, SN74S139 DECODERS /MULTIPLEXERS

- Designed Specifically for High-Speed:
  Memory Decoders
  Data Transmission Systems

- 'S138 and 'LS138 3-to-8-Line Decoders Incorporate 3 Enable Inputs to Simplify Cascading and/or Data Reception

- 'S139 and 'LS139A Contain Two Fully Independent 2-to-4-Line Decoders/ Demultiplexers

- Schottky Clamped for High Performance

TYPE	TYPICAL PROPAGATION DELAY (3 LEVELS OF LOGIC)	TYPICAL POWER DISSIPATION
'LS138	22 ns	32 mW
'S138	8 ns	245 mW
'LS139A	22 ns	34 mW
'S139	7.5 ns	300 mW

description

These Schottky-clamped TTL MSI circuits are designed to be used in high-performance memory-decoding or data-routing applications requiring very short propagation delay times. In high-performance memory systems these decoders can be used to minimize the effects of system decoding. When employed with high-speed memories utilizing a fast-enable circuit the delay times of these decoders and the enable time of the memory are usually less than the typical access time of the memory. This means that the effective system delay introduced by the Schottky-clamped system decoder is negligible.

The 'LS138 and 'S138 decode one-of-eight lines dependent on the conditions at the three binary select inputs and the three enable inputs. Two active-low and one active-high enable inputs reduce the need for external gates or inverters when expanding. A 24-line decoder can be implemented without external inverters and a 32-line decoder requires only one inverter. An enable input can be used as a data input for demultiplexing applications.

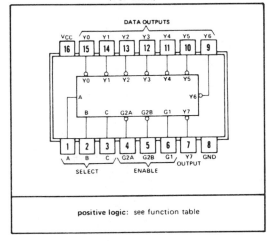

SN54LS138, SN54S138 . . . J OR W PACKAGE
SN74LS138, SN74S138 . . . J OR N PACKAGE
(TOP VIEW)

positive logic: see function table

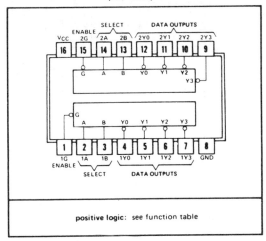

SN54LS139A, SN54S139 . . . J OR W PACKAGE
SN74LS139A, SN74S139 . . . J OR N PACKAGE
(TOP VIEW)

positive logic: see function table

The 'LS139A and 'S139 comprise two individual two-line-to-four-line-decoders in a single package. The active-low enable input can be used as a data line in demultiplexing applications.

All of these decoders/demultiplexers feature fully buffered inputs each of which represents only one normalized Series 54LS/74SL load ('LS138, 'LS139A) or one normalized Series 54S/74S load ('S138, 'S139) to its driving circuit. All inputs are clamped with high-performance Schottky diodes to suppress line-ringing and simplify system design. Series 54LS and 54S devices are characterized for operatio over the full military temperature range of −55°C to 125°C; Series 74LS and 74S devices are characterized for 0°C to 70°C industrial systems.

Figure 1.21(a)–(c) 74LS138 and 74LS139 data sheets (Courtesy Texas Instruments)

TYPES SN54LS138, SN54S138, SN54LS139A, SN54S139
SN74LS138, SN74S138, SN74LS139A, SN74S139
DECODERS/DEMULTIPLEXERS

functional block diagrams and logic

'LS138, 'S138

'LS138, 'S138 FUNCTION TABLE

INPUTS				OUTPUTS							
ENABLE		SELECT									
G1	G2*	C	B	A	Y0	Y1	Y2	Y3	Y4	Y5	Y6	Y7
X	H	X	X	X	H	H	H	H	H	H	H	H
L	X	X	X	X	H	H	H	H	H	H	H	H
H	L	L	L	L	L	H	H	H	H	H	H	H
H	L	L	L	H	H	L	H	H	H	H	H	H
H	L	L	H	L	H	H	L	H	H	H	H	H
H	L	L	H	H	H	H	H	L	H	H	H	H
H	L	H	L	L	H	H	H	H	L	H	H	H
H	L	H	L	H	H	H	H	H	H	L	H	H
H	L	H	H	L	H	H	H	H	H	H	L	H
H	L	H	H	H	H	H	H	H	H	H	H	L

*G2 = G2A + G2B

H = high level, L = low level, X = irrelevant

'LS139A, 'S139

'LS139A, 'S139 (EACH DECODER/DEMULTIPLEXER) FUNCTION TABLE

INPUTS			OUTPUTS			
ENABLE	SELECT					
G	B	A	Y0	Y1	Y2	Y3
H	X	X	H	H	H	H
L	L	L	L	H	H	H
L	L	H	H	L	H	H
L	H	L	H	H	L	H
L	H	H	H	H	H	L

H = high level, L = low level, X = irrelevant

schematics of inputs and outputs

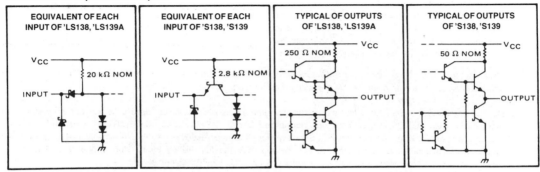

Figure 1.21(b)

TYPES SN54LS138, SN54LS139A, SN74LS138, SN74LS139A
DECODERS/DEMULTIPLEXERS

absolute maximum ratings over operating free-air temperature range (unless otherwise noted)

Supply voltage, V_{CC} (see Note 1)	7 V
Input voltage	7 V
Operating free-air temperature range: SN54LS138, SN54LS139A Circuits	$-55°C$ to $125°C$
SN74LS138, SN74LS139A Circuits	$0°C$ to $70°C$
Storage temperature range	$-65°C$ to $150°C$

NOTE 1: Voltage values are with respect to network ground terminal.

recommended operating conditions

	SN54LS138 SN54LS139A			SN74LS138 SN74LS139A			UNIT
	MIN	NOM	MAX	MIN	NOM	MAX	
Supply voltage, V_{CC}	4.5	5	5.5	4.75	5	5.25	V
High-level output current, I_{OH}			−400			−400	μA
Low-level output current, I_{OL}			4			8	mA
Operating free-air temperature, T_A	−55		125	0		70	°C

electrical characteristics over recommended operating free-air temperature range (unless otherwise noted)

PARAMETER	TEST CONDITIONS[†]		SN54LS138 SN54LS139A			SN74LS138 SN74LS139A			UNIT
			MIN	TYP[‡]	MAX	MIN	TYP[‡]	MAX	
V_{IH} High-level input voltage			2			2			V
V_{IL} Low-level input voltage					0.7			0.8	V
V_{IK} Input clamp voltage	V_{CC} = MIN, I_I = −18 mA				−1.5			−1.5	V
V_{OH} High-level output voltage	V_{CC} = MIN, V_{IH} = 2 V, $V_{IL} = V_{IL\,max}$, I_{OH} = −400 μA		2.5	3.4		2.7	3.4		V
V_{OL} Low-level output voltage	V_{CC} = MIN, V_{IH} = 2 V, $V_{IL} = V_{IL\,max}$	I_{OL} = 4 mA		0.25	0.4		0.25	0.4	V
		I_{OL} = 8 mA					0.35	0.5	
I_I Input current at maximum input voltage	V_{CC} = MAX, V_I = 7 V				0.1			0.1	mA
I_{IH} High-level input current	V_{CC} = MAX, V_I = 2.7 V				20			20	μA
I_{IL} Low-level input current	V_{CC} = MAX, V_I = 0.4 V				−0.4			−0.4	mA
I_{OS} Short circuit output current §	V_{CC} = MAX	'LS138	−20		−100	−20		−100	mA
		'LS139A	−20		−100	−20		−100	
I_{OS} Supply current	V_{CC} = MAX, Outputs enabled and open	'LS138		6.3	10		6.3	10	mA
		'LS139A		6.8	11		6.8	11	

[†] For conditions shown as MIN or MAX, use the appropriate value specified under recommended operating conditions for the applicable device type.
[‡] All typical values are at V_{CC} = 5 V, T_A = 25°C.
§ Not more than one output should be shorted at a time.

switching characteristics, V_{CC} = 5 V, T_A = 25°C

PARAMETER[¶]	FROM (INPUT)	TO (OUTPUT)	LEVELS OF DELAY	TEST CONDITIONS	SN54LS138 SN74LS138			SN54LS139A SN74LS139A			UNIT
					MIN	TYP	MAX	MIN	TYP	MAX	
t_{PLH}	Binary Select	Any	2			13	20		13	20	ns
t_{PHL}						27	41		22	33	ns
t_{PLH}			3	C_L = 15 pF, R_L = 2 kΩ, See Note 2		18	27		18	29	ns
t_{PHL}						26	39		25	38	ns
t_{PLH}	Enable	Any	2			12	18		16	24	ns
t_{PHL}						21	32		21	32	ns
t_{PLH}			3			17	26				ns
t_{PHL}						25	38				ns

[¶] t_{PLH} ≡ propagation delay time, low-to-high-level output; t_{PHL} ≡ propagation delay time, high-to-low-level output.
NOTE 2: Load circuits and waveforms are shown on page 3-11.

Figure 1.21 (c)

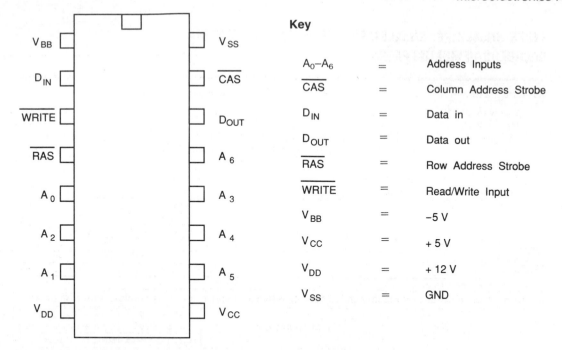

Figure 1.22   Typical 16K × 1 memory chip

Key

A_0–A_6	=	Address Inputs
$\overline{CAS}$	=	Column Address Strobe
D_{IN}	=	Data in
D_{OUT}	=	Data out
$\overline{RAS}$	=	Row Address Strobe
$\overline{WRITE}$	=	Read/Write Input
V_{BB}	=	–5 V
V_{CC}	=	+ 5 V
V_{DD}	=	+ 12 V
V_{SS}	=	GND

required within each memory device. Dynamic memory has provided a method of increasing the packing density within each chip, such that it is possible to have, for example 16K × 1, 64K × 1 and 256K × 1 dynamic memory devices readily available.

There are some additional complications, however, since dynamic RAM must be **refreshed**, and also, its memory addresses must be **multiplexed**. This means that, for example in a 16K × 1 chip, the data for its 14 address bits must be provided as two sequential groups of 7. This means that devices known as multiplexers are likely to be found in computer circuits that employ dynamic memory chips.

Internally, all memory chips have a matrix of memory cells set out in **rows** and **columns**. When addressing dynamic RAM, the data to select the **row** and **column** of the memory array is supplied separately, under the control of two signals known as $\overline{RAS}$ and $\overline{CAS}$ (Row Address Strobe and Column Address Strobe). Some timing logic must be provided to control the multiplexing operation so that the $\overline{RAS}$ and $\overline{CAS}$ signals are

applied to the memory array at exactly the right time.

The pin connections of a typical 16K × 1 memory device as shown in *Figure 1.22*. The diagram shows only seven address lines, but these are used for all 14 address bits. There is a separate data input and data output pin which allows very fast READ–MODIFY–WRITE operation. Power requirements are such that +12 V, +5 V, −5 V and GND are required for normal operation.

Address bit multiplexing can be achieved with a pair of 74 Series multiplexer chips such as 74LS157's which are enabled at the appropriate moment by the timing signals. These effectively combine the 14 address lines into 7 with the appropriate timing. These timing signals may be provided in one of a number of ways, such as with monostables, with a digital delay line or with logic devices.

An example of a dynamic memory interface that uses logic devices to derive the timing signals is shown in *Figure 1.23*. The complete circuit is included to provide an idea of the added complexity which dynamic memory brings to a system.

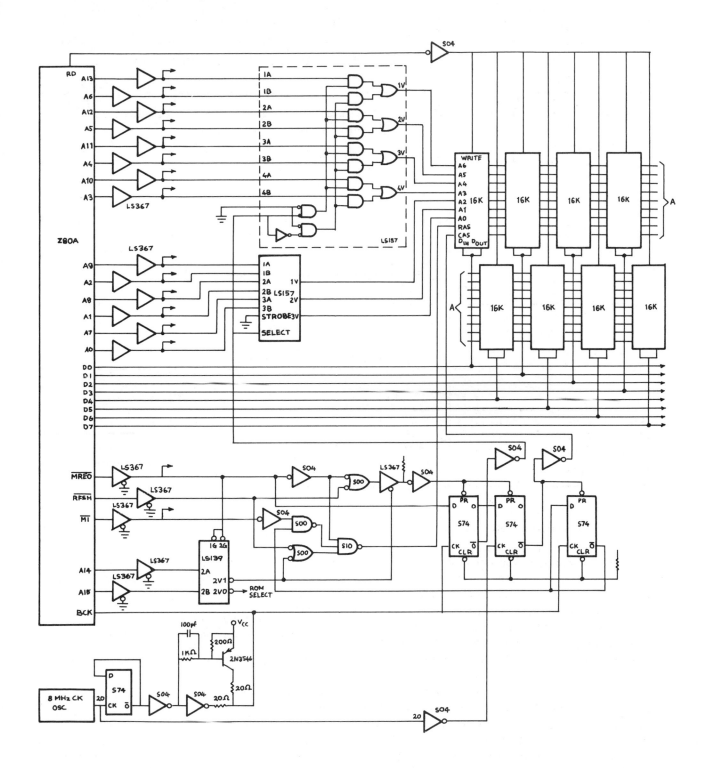

Figure 1.23 Z80 16K dynamic RAM interface (Courtesy Mostek)

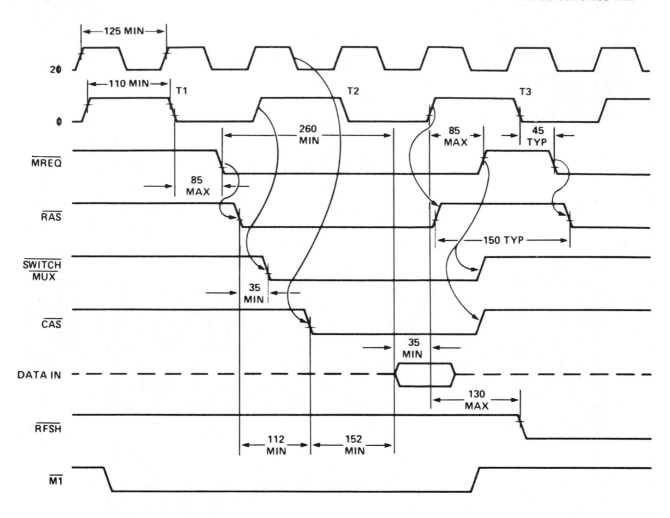

Figure 1.24 Z80 dynamic RAM timing

The system timing diagram is given in *Figure 1.24.* It can be seen that the $\overline{MREQ}$ signal triggers the $\overline{RAS}$ signal which clocks the ROW address data into the memory chip. Shortly after, the multiplexer is switched and then the $\overline{CAS}$ signal becomes active. This then clocks the COLUMN address into the device. On the rising edge of T_2, data is READ into memory.

Summary

The main points covered in this chapter relate to the types of integrated circuit found in computer systems and those that are particularly found in the address bus circuits.

- Most microcomputer circuits employ both TTL and NMOS technology while some low power-systems employ CMOS.

- Address decoders are used to select between memory devices and input/output ports.

- Address bus buffers increase the fan-out of the CPU.

- Full address decoding is preferred in general since it does not create 'mirror' addresses in memory.

- Dynamic memory devices generally require complex refresh circuits and use multiplexers to switch address lines.

Questions

1.1 Briefly describe the main differences between a CMOS gate and a TTL gate.

1.2 Under what circumstances may a designer use a 74 Series chip in preference to a 74LS Series chip?

1.3 Which octal bus buffer would be best to use if non- inverting outputs were required in a system to be used in arctic conditions?

1.4 What is the maximum current that could be expected from a 74LS240 with an output short circuited to the 0 volt line?

1.5 What is the typical value of the hysteresis voltage the 74LS240/44 gates provide?

1.6 A small microcomputer system has three memory devices, a 4K $\times$ 8 ROM and two 2K $\times$ 8 RAM chips. It is required to have them at addresses 0000 hex, 2000 hex and 2800 hex respectively.
Draw the circuit diagram for the decoding part of the system.

1.7 Write down the lowest and highest memory addresses in each of eight 4K $\times$ 8 ROM chips whose **enable** inputs are connected to the outputs of a 3–8 line decoder with inputs connected to address lines A_{12}, A_{13} and A_{14}.

1.8 Briefly describe the advantages and disadvantages of **partial addressing decoding**.

1.9 In what circumstances would **linear** address decoding be used in a microcomputer system?

1.10 Give two advantages of using memory mapped input/output in a microprocessor system.

1.11 A decoder has two inputs and four outputs. It has active **low** outputs and an active **low enable** input. Write down its **truth table**.

1.12 List the advantages and disadvantages of using 'dynamic' memory in a microcomputer system.

1.13 What are the main characteristics of a 74LS139 chip?

1.14 Why are some microprocessors forced to use memory mapped input/output?

TWO

Data and control bus circuits

When you have completed this chapter, you should be able to:

1. Explain the need for tri-state circuits on a micro-computer data bus.
2. Describe the operation of the 'chip enable' signals in a system.
3. Appreciate the need for bi-directional buffers in data-bus circuits.
4. Understand the operation of a typical micropro-cessor clock circuit.
5. Relate the operation of a typical microprocessor instruction to the corresponding timing diagram and the control signals.
6. Explain the function and operation of a simple practical microcomputer circuit.

2.1 CIRCUITS CONNECTED TO THE DATA BUS

In a microcomputer system the three system buses have different characteristics. The address bus is **unidirectional**, which means that all the information goes from the CPU to the rest of the system. The control bus operates in **two direc-tions**, which means that some signals are **inputs** to the CPU and some signals are **outputs** from the CPU. The data bus is said to be **bidirectional**. This means that the same pins on the CPU and all the other devices connected to the data bus can carry data in both directions. The same pins are used for input data and output data.

Consider *Figure 2.1* which shows the data bus connected to a CPU, a memory device and an input/output (I/O) port. Under most normal cir-cumstances, the CPU will be either **reading** data from one of the other devices or **writing** data to one of them.

If the CPU is reading data from the input port, the memory device is not involved in the data transfer and its connections must therefore not interfere with the input operation. It is desirable to be able to effectively disconnect the memory from the data bus at that moment.

Similarly, when memory operations are taking place it is desirable to be able to disconnect the I/O device effectively.

The problem arises not when another input is connected to the bus, but when two **outputs** are simultaneously connected. If the CPU memory and I/O port are all NMOS devices, the circuits connected together could be as shown in *Figure 2.2*.

Imagine that the CPU has attempted to output a logic 1 on to the data line to be read by the memory device. However, the previous input from the port was a logic 0 which is still active on the port output pin. The 1s and 0s are established by each device turning ON one of its output tran-sistors. In this case, current would flow from the +5 V supply, through the top transistor in the CPU output, along the data bus line and through the lower I/O transistor to 0 V. This would effectively place two conducting transistors across the +5 V supply with disastrous consequences.

The only way to ensure that two outputs never attempt to place **conflicting** data on the bus is to

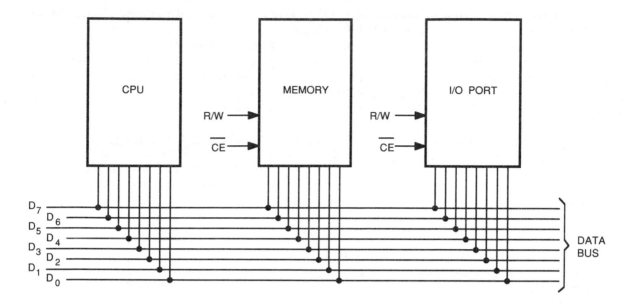

Figure 2.1   Data bus connections

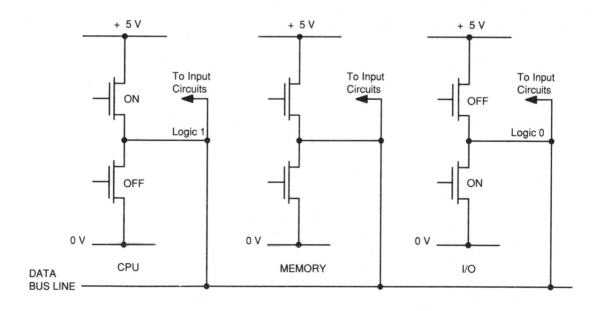

Figure 2.2   Outputs connected to the data bus

ensure that all unused outputs adopt a **high-impedance** state. This is known as the **tri-state** output – i.e. the logic device now has three output states, logic 1, logic 0 and high impedance.

In *Figure 2.2*, if the unused output from the I/O port had **both** transistors turned **off**, then it would be unable to affect the state of the bus line and there would be no conflict.

So, the rules for the operation of the data bus are as follows:

(a) ONE device acts as an **output** to write the bus data.
(b) ONE device acts as an **input** to read the bus data.
(c) ALL other devices must be in their **tri-state** condition which effectively disconnects them from the bus.

Logic circuits are now available, whether TTL, CMOS or NMOS which have 3-state outputs. Without this capability they cannot be connected to the data bus in a system. Normally all circuits intended for use in microprocessor systems such as memories or I/O ports have 3-state outputs built in.

In all of these devices, the condition of the 3-state output or the corresponding input, is determined by the control bus signals or those derived from them, together with an address signal.

Signals that are derived from the **address** bus are used to choose which chip within a system is to be active at any instant. These are known as **chip select** signals, normally shown as $\overline{\text{CS}}$.

Signals that indicate whether a chip will read or write, or whether it will act as memory or an I/O port are derived from the control bus. They are generally known as **chip enable** signals, shown as $\overline{\text{CE}}$.

Unfortunately many books and manufacturers confuse these basic definitions. Both types of signal are necessary for the correct operation of a system.

2.2 CHIP ENABLE SIGNALS

The control bus outputs of most microprocessors contain signals that allow the system to select between **read** and **write** operations and between **memory** and **input/output**. In the Z80, the signals are:

$\overline{\text{RD}}$ – Read
$\overline{\text{WR}}$ – Write
$\overline{\text{IORQ}}$ – Input/output request
$\overline{\text{MREQ}}$ – Memory request

These are combined in different ways to provide the required chip enable signals for each part of the system.

For example, an EPROM in a system could be **enabled** by a combination of $\overline{\text{RD}}$ and $\overline{\text{MREQ}}$. An input port may require $\overline{\text{RD}}$ and $\overline{\text{IORQ}}$. An output port may require $\overline{\text{WR}}$ and $\overline{\text{IORQ}}$.

Since the majority of inputs are **active low** (activated by a logic 0), the gates used for these systems could be as shown in *Figure 2.3*.

In some manufacturers designs, the chip enable signal is combined with the **chip select** signal before it is sent to the device. This is quite common for memory devices; typically, the chip enable signal is used to activate the **address decoder**. However, an example of this type of circuit is shown in *Figure 2.4*.

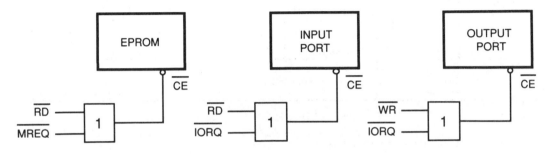

Figure 2.3 Chip enable signals

Combining CE and CS in one signal

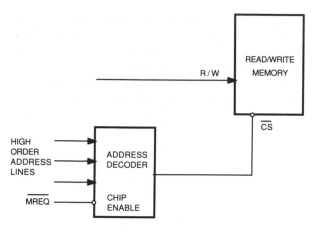

Figure 2.4   Combining $\overline{CE}$ and $\overline{CS}$ in one signal

Notice that the $\overline{CE}$ input of the address decoder chip shown has an 'active low' input and so it can be connected directly to the $\overline{MREQ}$ line of the processor. The **read/write** selection may be accomplished either by combining the processor read and write lines, or by simply using only the $\overline{WR}$ line and assuming that all operations which are not writes must be reads.

The logic signal output from the circuit shown in *Figure 2.4* can be derived by examining the timing diagram for the op-code fetch or memory read cycles (*Figure 2.5*). It assumes that there is a valid memory address on the address bus for the chip in question. Note that the memory device cannot be selected during the **refresh** cycle since the refresh address is only placed on the lower 7

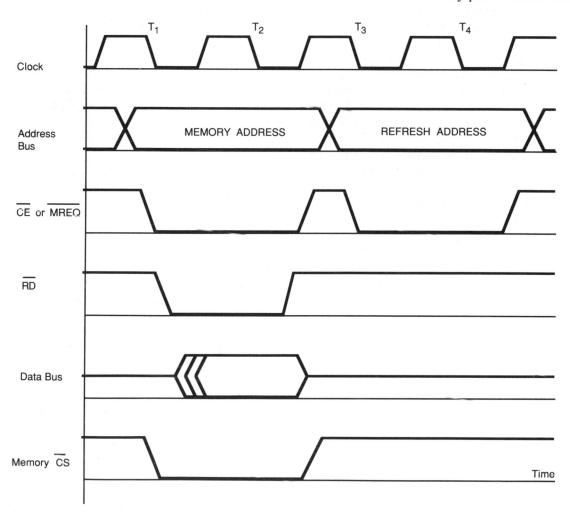

Figure 2.5   Z80 op-code fetch timing

bits of the address bus (A_0–A_6) and these would not normally be used as the inputs to the address decoder shown. The apparent double waveform for the address bus is used to indicate that some address lines may be at a logic 1 and others at a logic 0 during the machine cycle. The multiple lines at the beginning of the data bus waveform indicate that some memory devices may provide their data more quickly than others, and the exact time is therefore uncertain.

2.3 DATA BUS BUFFERS

In large systems, it is a common design technique to provide a buffer chip in the data bus lines. There are several reasons for this. First, it allows the drive capability of the microprocessor to be considerably increased which means that it can supply sufficient current for a bus that may have a number of devices connected to it. This means that the system, however large, will not load the CPU. Secondly, it provides a means of isolation, so that if required, the CPU and associated circuitry may be effectively disconnected from the rest of the system.

The data bus must allow signals to pass in both directions, and therefore a **bidirectional** buffer is required. This consists of two tri-state buffers, wired 'back-to-back', with appropriate control logic. A typical circuit is shown in *Figure 2.6*.

Two control signals are required. The enable signal is provided so that the whole device can be either enabled or disabled as required. When disabled, both outputs assume their high-impedance

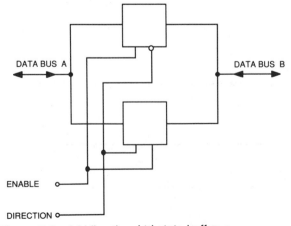

Figure 2.6 A bidirectional tri-state buffer

condition. When the device is enabled, the buffer that operates is determined by the **direction** input. A logic 1 enables the lower buffer, while a logic 0 enables the upper buffer.

Various manufacturers supply bidirectional bus buffers, each with a range of options, such as:

- Inverting or non-inverting.
- Active high or active low enable.
- Tri-state or open collector outputs.

These devices are also known as **bus transceivers**. A typical device in the 74 Series is the 74LS245, which is known as an **octal bus transceiver** with **3-state outputs**.

The data sheet in *Figure 2.7* (pages 35–6) is taken from the Texas Instruments TTL Data Book. Notice from the data sheet that in this case the enable input is active low, so that when it is at a logic 1, the whole device is disabled.

The 20-pin device contains two sets of octal tri-state buffers wired 'back to back' with a pair of controlling gates which determine which set of buffers are enabled at any moment. Note that the drive capability of the buffer is not necessarily the same in both directions of operation.

2.4 THE OPERATION OF THE CONTROL BUS

The **control** bus is derived from the CPU control unit and contains a number of signals which are used to determine HOW the system will operate at any moment. Different microprocessors have different control signals, but they all perform more or less the same basic operations.

The majority of control bus signals are outputs from the CPU. They generally have tri-state outputs so that under certain circumstances they can be switched off altogether. However, some control bus signals are **CPU inputs**. This means that they are derived elsewhere in the system and may be used to control the CPU operation in some way.

The control signals of the Z80 CPU are conveniently divided into three main groups: system control, CPU control and CPU bus control. In addition to these groups however, there is the one signal without which nothing at all would happen – the **clock** input. It is this signal that is considered first.

TTL
MSI

TYPES SN54LS245, SN74LS245
OCTAL BUS TRANSCEIVERS WITH 3-STATE OUTPUTS

- Bi-directional Bus Transceiver in a High-Density 20-Pin Package

- 3-State Outputs Drive Bus Lines Directly

- P-N-P Inputs Reduce D-C Loading on Bus Lines

- Hysteresis at Bus Inputs Improve Noise Margins

- Typical Propagation Delay Times, Port-to-Port . . . 8 ns

- Typical Enable/Disable Times . . . 17 ns

TYPE	I_{OL} (SINK CURRENT)	I_{OH} (SOURCE CURRENT)
SN54LS245	12 mA	−12 mA
SN74LS245	24 mA	−15 mA

logic diagram (positive logic)   logic symbol

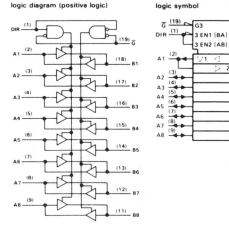

description

These octal bus transceivers are designed for asynchronous two-way communication between data buses. The control function implementation minimizes external timing requirements.

The device allows data transmission from the A bus to the B bus or from the B bus to the A bus depending upon the logic level at the direction control (DIR) input. The enable input ($\overline{G}$) can be used to disable the device so that the buses are effectively isolated.

The SN54LS245 is characterized for operation over the full military temperature range of −55°C to 125°C. The SN74LS245 is characterized for operation from 0°C to 70°C.

schematics of inputs and outputs

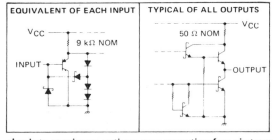

FUNCTION TABLE

ENABLE $\overline{G}$	DIRECTION CONTROL DIR	OPERATION
L	L	B data to A bus
L	H	A data to B bus
H	X	Isolation

H = high level, L = low level, X = irrelevant

absolute maximum ratings over operating free-air temperature range (unless otherwise noted)

Supply voltage, V_{CC} (see Note 1) . . . . 7 V
Input voltage . . . . 7 V
Operating free-air temperature range: SN54LS245 . . . . −55°C to 125°C
                                        SN74LS245 . . . . 0°C to 70°C
Storage temperature range . . . . −65°C to 150°C
Off state output voltage . . . . 5·5V

NOTE 1: Voltage values are with respect to network ground terminal.

Figure 2.7(a)–(b) 74LS245 data sheets (Courtesy Texas Instruments)

TYPES SN54LS245, SN74LS245
OCTAL BUS TRANSCEIVERS WITH 3-STATE OUTPUTS

REVISED FEBRUARY 1979

recommended operating conditions

PARAMETER	SN54LS245 MIN	SN54LS245 NOM	SN54LS245 MAX	SN74LS245 MIN	SN74LS245 NOM	SN74LS245 MAX	UNIT
Supply voltage, V_{CC}	4.5	5	5.5	4.75	5	5.25	V
High-level output current, I_{OH}			−12			−15	mA
Low-level output current, I_{OL}			12			24	mA
Operating free-air temperature, T_A	−55		125	0		70	°C

electrical characteristics over recommended operating free-air temperature range (unless otherwise noted)

PARAMETER		TEST CONDITIONS[†]		SN54LS245 MIN	SN54LS245 TYP[‡]	SN54LS245 MAX	SN74LS245 MIN	SN74LS245 TYP[‡]	SN74LS245 MAX	UNIT
V_{IH}	High-level input voltage			2			2			V
V_{IL}	Low-level input voltage					0.7			0.8	V
V_{IK}	Input clamp voltage	V_{CC} = MIN,	I_I = −18 mA			−1.5			−1.5	V
	Hysteresis $(V_{T+} - V_{T-})$ A or B input	V_{CC} = MIN		0.2	0.4		0.2	0.4		V
V_{OH}	High-level output voltage	V_{CC} = MIN, V_{IH} = 2 V, $V_{IL} = V_{IL}$ max	I_{OH} = −3 mA	2.4	3.4		2.4	3.4		V
			I_{OH} = MAX	2			2			
V_{OL}	Low-level output voltage	V_{CC} = MIN, V_{IH} = 2 V, $V_{IL} = V_{IL}$ max	I_{OL} = 12 mA			0.4			0.4	V
			I_{OL} = 24 mA						0.5	
I_{OZH}	Off-state output current, high-level voltage applied	V_{CC} = MAX, $\overline{G}$ at 2 V	V_O = 2.7 V			20			20	µA
I_{OZL}	Off-state output current, low-level voltage applied		V_O = 0.4 V			−200			−200	
I_I	Input current at	A or B	V_{CC} = MAX,	V_I = 5.5 V			0.1			0.1	mA
	maximum input voltage	DIR or $\overline{G}$		V_I = 7 V			0.1			0.1	
I_{IH}	High-level input current	V_{CC} = MAX,	V_{IH} = 2.7 V			20			20	µA
I_{IL}	Low-level input current	V_{CC} = MAX,	V_{IL} = 0.4 V			−0.2			−0.2	mA
I_{OS}	Short-circuit output current[¶]	V_{CC} = MAX		−40		−225	−40		−225	mA
I_{CC}	Supply current — Total, outputs high	V_{CC} = MAX, Outputs open			48	70		48	70	mA
	Total, outputs low				62	90		62	90	
	Outputs at Hi-Z				64	95		64	95	

[†]For conditions shown as MIN or MAX, use the appropriate value specified under recommended operating conditions.
[‡]All typical values are at V_{CC} = 5 V, T_A = 25°C.
[¶]Not more than one output should be shorted at a time, and duration of the short-circuit should not exceed one second.

switching characteristics, V_{CC} = 5 V, T_A = 25°C

PARAMETER		TEST CONDITIONS				MIN	TYP	MAX	UNIT
t_{PLH}	Propagation delay time, low-to-high-level output	C_L = 45 pF,	R_L = 667 Ω,	See Note 2			8	12	ns
t_{PHL}	Propagation delay time, high-to-low-level output						8	12	ns
t_{PZL}	Output enable time to low level						27	40	ns
t_{PZH}	Output enable time to high level						25	40	ns
t_{PLZ}	Output disable time from low level	C_L = 5 pF,	R_L = 667 Ω,	See Note 2			15	25	ns
t_{PHZ}	Output disable time from high level						15	25	ns

NOTE 2: Load circuit and waveforms are shown on page 1-15.

Figure 2.7(b)

2.5 THE CLOCK

Typical CPU clock speeds can vary between 1 MHz and 30 MHz. Generally the **clock** is a signal derived from a stable crystal oscillator which is built with some high speed inverters and some timing components.

For example, a Z80 requires a simple **single-phase** clock which is a square wave of up to 8 MHz. Other processors, however, require a **multi-phase** clock (*Figure 2.8*). These require two or more pulsed waveforms which are out of phase with each other. Up to four phases are required with some processors.

The simplest clock waveform to generate is the single-phase clock with a +5 V amplitude, and this can be produced in many ways. Transistor oscillators with crystal control could provide the necessary stability, but a better waveform would normally be obtained directly from an oscillator made of a pair of inverting gates.

Generally an oscillator frequency of **twice** the clock frequency is chosen so that the resulting waveform can be divided by two, thus providing a square waveform with added buffering. *Figure 2.9* shows a typical oscillator clock circuit.

Such a clock can be used for most microprocessors that require a single-phase clock. However, for higher speed operation, and to improve the rise time of the clock signal, the Z80 manufacturers recommend an additional pull-up circuit (*Figure 2.10*).

The main requirements of a CLOCK waveform are:

(a) Fast rise and fall times.
(b) Stable frequency.
(c) Clean waveform – noise free.

As long as these requirements can be met, the design of the clock circuit is generally not critical, so many different designs will be seen in different systems. More complex clock requirements often demand the use of a dedicated clock chip.

2.6 MAIN CONTROL BUS SIGNALS

The main Z80 control bus signals are shown diagrammatically in *Figure 2.11* (overleaf). They are conveniently subdivided into three groups: the

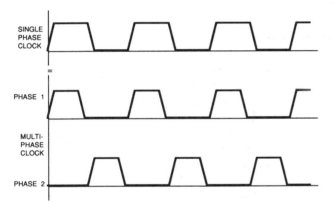

Figure 2.8 Single and multi-phase clocks

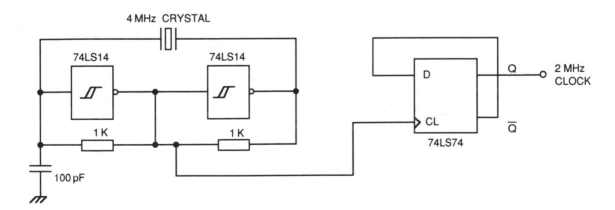

Figure 2.9 Typical microprocessor clock circuit

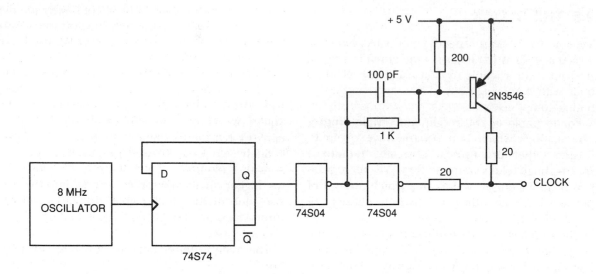

Figure 2.10   Recommended Z80 clock circuit (4 MHz)

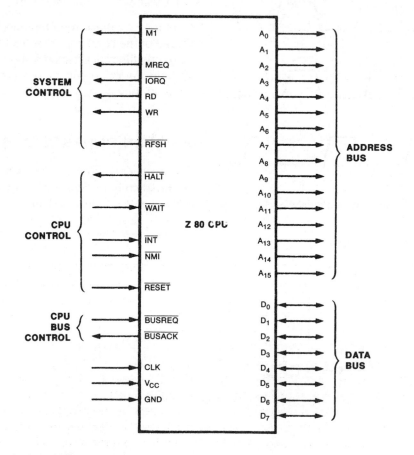

Figure 2.11   Z80 pin description

system control group, the CPU control group, and the CPU bus control group. Each of these groups has a particular function so it will be convenient to consider them separately.

System Control Signals

All the system control signals are outputs from the CPU and are all active low. Four of the six signals are regarded as the main control bus signals, i.e. **memory request**, **input/output request**, **read**, and **write**. In addition, there are two other signals peculiar to the Z80, the **refresh** signal and the **machine cycle 1** signal.

$\overline{\text{M1}}$ – Machine cycle 1 (output, active low)

$\overline{\text{M1}}$, as the name suggests, is a signal that is active only during the first machine cycle of each instruction execution. $\overline{\text{M1}}$, together with $\overline{\text{MREQ}}$, indicates that the current machine cycle is an op-code fetch cycle. $\overline{\text{M1}}$, together with $\overline{\text{IORQ}}$, indicates an interrupt–acknowledge cycle.

$\overline{\text{RFSH}}$ – Refresh (output, active low)

$\overline{\text{RFSH}}$ is signal unique to the Z80. Together with the $\overline{\text{MREQ}}$, it indicates that the lower 7 bits of the system's address bus can be used as a refresh address to the system's dynamic memory. If the system has no dynamic memory devices, the refresh signal is clearly redundant and is simply left disconnected.

$\overline{\text{MREQ}}$ – Memory request (output, active low, 3 state)

Memory request is one of the main processor control signals which controls the operation of memory devices within the system. $\overline{\text{MREQ}}$ indicates that the address bus holds a valid address for a memory read or memory write operation. It is not until the $\overline{\text{MREQ}}$ signal is active that the address on the address bus can be regarded as being stable.

$\overline{\text{IORQ}}$ – Input/output request (output, active low, 3 state)

Input/output request indicates that the lower half of the address bus holds a valid I/O address for an I/O read or write operation. $\overline{\text{IORQ}}$ is also generated with $\overline{\text{M1}}$ during an interrupt–acknowledge cycle to indicate that an interrupt response vector has been placed on the data bus.

$\overline{\text{RD}}$ – Read (output, active low, 3 state)

$\overline{\text{RD}}$ indicates that the CPU wants to read data from memory or an I/O device. The addressed I/O device or memory should use this signal to gate data onto the CPU data bus.

$\overline{\text{WR}}$ – Write (output, active low, 3 state)

$\overline{\text{WR}}$ indicates that the CPU data bus holds valid data to be stored at the memory address or I/O location indicated by the address bus.

With some peripherals the $\overline{\text{WR}}$ line may be used for both read and write. When the WR signal is **high** it can be taken as a **read** signal but when it is **low** it is taken as a **write** signal.

For simple systems, the CPU is the only device that can control the system buses, and therefore the control bus signals are generated by the CPU. However, in larger systems, other computer chips such as **DMA** controllers (Direct Memory Access) may also control the system buses. In such cases the system control signals may be forced into their inactive state and control of the buses may be taken over by the DMA controller chip. This is referred to later under CPU bus control.

CPU Control Signals

The next group of control signals are those that are used to control the CPU. They are generally generated by other peripheral devices or by external circuits. Of the five signals there are four inputs and one output. The only output is the HALT signal.

$\overline{\text{HLT}}$ – Halt (output, active low)

$\overline{\text{HLT}}$ indicates that the CPU has executed a **HALT** instruction and is awaiting an interrupt signal before it can resume operation. When halted, the CPU executes NOPs to maintain memory refresh. Once the Z80 has executed HALT instruction the only way it can be reactivated is by a logic input to one of the three input pins in CPU control group.

$\overline{\text{RESET}}$ – Reset (input, active low)

$\overline{\text{RESET}}$ initialises the CPU as follows.

It resets the interrupt enable flip-flop, clears the program counter and registers I and R, and sets the interrupt status to mode 0. During reset time, the address and data bus goes to a high-impedance state, and all control output signals go to the inactive state. Note that reset must be active for a minimum of three full clock cycles before the reset operation is complete. Generally the reset signal is derived from a switch somewhere in the system. This acts as an emergency restoration system since it overrides whatever the processor is doing at any time and can therefore rescue programs that have gone wrong.

$\overline{\text{WAIT}}$ – Wait (input, active low) $\overline{\text{WAIT}}$ indicates to the CPU that the addressed memory or I/O devices are not ready for a data transfer. The CPU continues to enter a **wait** state as long as this signal is active. Extended wait periods can prevent the CPU from refreshing dynamic memory properly. Every time the wait line is checked, and found to be active, an extra clock cycle is added to the current machine cycle.

$\overline{\text{INT}}$ – Interrupt request (input, active low)
Interrupt request is generated by I/O devices. The CPU honours a request at the end of the current instruction if the internal software control interrupt enable flip-flop (IFF) is enabled. INT is normally wire-ORed and requires an external pull-up resistor for its correct operation. Normally in systems that contain more than one device that may generate an interrupt, they both share the single interrupt pin, but some form of priority system must be established with external logic so that one device has precedence over the other if they both interrupt simultaneously.

$\overline{\text{NMI}}$ – Non-maskable interrupt (input, negative edge trigger) $\overline{\text{NMI}}$ has a higher priority than INT. NMI is always recognised at the end of the current instruction, independent of the status of the interrupt enable flip-flop, and automatically forces the CPU to restart at location 0066H. If the non-maskable interrupt pin is used, it is the responsibility of the programmer to provide the required software at location 0066H to generate the appropriate response to the signal.

CPU Bus Control

There are two signals, one input and one output, that control the operation of the system buses. Under normal circumstances the CPU has com-

plete control of address data and control buses, but when other chips are designed into the computer system which may also require control of the buses, then some form of bus sharing logic must be implemented. This is embodied in the bus request and bus acknowledge control signals. If an external circuit requires to control the buses, then it sends a bus request signal to the CPU. The CPU then generates a bus acknowledge signal and relinquishes control to the external circuit.

$\overline{\text{BUSREQ}}$ – Bus request (input, active low)
Bus request has a higher priority than $\overline{\text{NMI}}$ and is always recognised at the end of the current machine cycle. $\overline{\text{BUSREQ}}$ forces the CPU address bus, data bus and control signals $\overline{\text{MREQ}}$, $\overline{\text{IORQ}}$, $\overline{\text{RD}}$ and $\overline{\text{WR}}$ to go to a high-impedance state so that other devices can control these lines. $\overline{\text{BUSREQ}}$ is normally wired-ORed and requires an external pull-up resistor for these applications. Extended BUSREQ periods due to extensive direct memory access operations can prevent the CPU from properly refreshing dynamic RAMs.

$\overline{\text{BUSACK}}$ – Bus acknowledge (output, active low)   Bus acknowledge indicates to the requesting device that the CPU address bus, data bus and control signals $\overline{\text{MREQ}}$, $\overline{\text{IORQ}}$, $\overline{\text{RD}}$ and $\overline{\text{WR}}$ have entered their high impedance state. The external circuitry can now control these lines.

2.7 CPU TIMING

The Z80 CPU executes instructions by proceeding through a specific sequence of operations:

(a) Memory read or write.
(b) I/O device read or write.
(c) Interrupt acknowledge.

The timing waveforms associated with the operation of the Control bus have already been discussed in some detail in *Microelectronics NII*. However, the diagrams given there were slightly simplified and *Figures 2.12* and *2.13* are an indication of the complexity of the timing arrangements of a typical microprocessor. They are taken from a manufacturer's data book for the Z80.

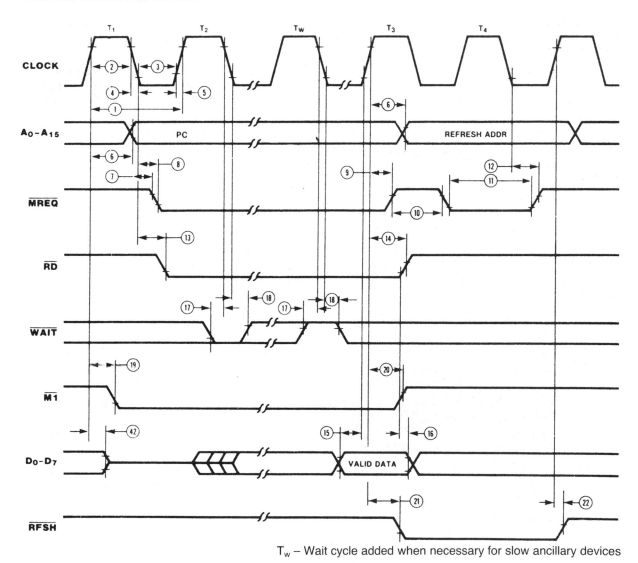

Figure 2.12 Z80 instruction op-code fetch timing (Courtesy SGS)

The details of every time marked in the figures are very precisely defined. All of the timing is taken from the moment at which a clock cycle reaches its logic 1 level in T_1. The timing chart gives typical figures for each part of the waveform for processors of different speeds. The Z8400 has a maximum clock speed of 2.5 MHz, the Z8400A runs at 4 MHz, Z8400B at 6 MHz, and the Z8400H at 8 MHz. Clearly most of the times are proportional to one another although as the speed of the clock increases some of the times do not decrease. This is because the processor requires a minimum time in which to operate no matter how fast the clock runs.

Figure 2.12 has some broken lines in the centre of it, which indicate that the cycles can be interrupted by a wait state or any number of wait states which are added between T_2 and T_3 while the WAIT input is held low. When the WAIT line returns to logic 1 the cycle continues with T_3 and T_4 as normal.

Note that the Z8400 is the SGS type number for the Z80. Another manufacturer refers to it as the MK 3881.

AC Characteristics

Number	Symbol	Parameter	Z84C00		Z84C00A		Z84C00B	
			Min (ns)	Max (ns)	Min (ns)	Max (ns)	Min (ns)	Max (ns)
1	TcC	Clock Cycle Time	400	DC	250	DC	165	DC
2	TwCh	Clock Pulse Width (High)	180	DC	110	DC	65	DC
3	TwCl	Clock Pulse Width (Low)	180	DC	110	DC	65	DC
4	TfC	Clock Fall Time	—	30	—	30	—	20
5	TrC	Clock Rise Time	—	30	—	30	—	20
6	TdCr(A)	Clock ↑ to Address Valid Delay	—	145	—	110	—	90
7	TdA(MREQf)	Address Valid to $\overline{MREQ}$ ↓ Delay	125	—	65	—	35	—
8	TdCf(MREQf)	Clock ↓ to $\overline{MREQ}$ ↓ Delay	—	100	—	85	—	70
9	TdCr(MREQr)	Clock ↑ to $\overline{MREQ}$ ↑ Delay	—	100	—	85	—	70
10	TwMREQh	$\overline{MREQ}$ Pulse Width (High)	170		110		65	
11	TwMREQl	$\overline{MREQ}$ Pulse Width (Low)	360	—	220	—	135	—
12	TdCf(MREQr)	Clock ↓ to $\overline{MREQ}$ ↑ Delay	—	100	—	85	—	70
13	TdCf(RDf)	Clock ↓ to $\overline{RD}$ ↓ Delay	—	130	—	95	—	80
14	TdCr(RDr)	Clock ↑ to $\overline{RD}$ ↑ Delay	—	100	—	85	—	70
15	TsD(Cr)	Data Setup Time to Clock ↑	50		35		30	
16	ThD(RDr)	Data Hold Time to $\overline{RD}$ ↑	0	—	0	—	0	—
17	TsWAIT(Cf)	$\overline{WAIT}$ Setup Time to Clock ↓	70	—	70	—	60	—
18*	ThWAIT(Cf)	$\overline{WAIT}$ Hold Time after Clock ↓	20	—	10	—	10	—
19	TdCr(Mlf)	Clock ↑ to $\overline{M1}$ ↓ Delay	—	130	—	100	—	80
20	TdCr(Mlr)	Clock ↑ to $\overline{M1}$ ↑ Delay		130		100		80
21	TdCr(RFSHf)	Clock ↑ to $\overline{RFSH}$ ↓ Dealy	—	180	—	130	—	110
22	TdCr(RFSHr)	Clock ↑ to $\overline{RFSH}$ ↑ Delay	—	150	—	120	—	100
23	TdCf(RDr)	Clock ↓ to $\overline{RD}$ ↑ Delay	—	110	—	85	—	70
24	TdCr(RDf)	Clock ↑ to $\overline{RD}$ ↓ Dealy	—	110	—	85	—	70
25	TsD(Cf)	Data Setup to Clock ↓ during M_2, M_3, M_4 or M_5 Cycles	60		50		40	
26	TdA(IORQf)	Address Stable prior to $\overline{IORQ}$ ↓	320	—	180	—	110	—
27	TdCr(IORQf)	Clock ↑ to $\overline{IORQ}$ ↓ Delay	—	100*	—	75	—	65
28	TdCf(IORQr)	Clock ↓ to $\overline{IORQ}$ ↑ Delay	—	110	—	85	—	70
29	TdCf(WRf)	Data Stable prior to $\overline{WR}$ ↓	190	—	80	—	25	—
30	TdDf(WRf)	Clock ↓ to $\overline{WR}$ ↓ Delay		90		80		70
31	TwWR	$\overline{WR}$ Pulse Width	360	—	220	—	135	—
32	TdCf(WRr)	Clock ↓ to $\overline{WR}$ ↑ Delay	—	100	—	80	—	70
33	TdD(WRf)	Data Stable prior to $\overline{WR}$ ↓	20	—	−10	—	−55	—
34	TdCr(WRf)	Clock ↑ to $\overline{WR}$ ↓ Delay	—	100*	—	65	—	60
35	TdWRr(D)	Data Stable from $\overline{WR}$ ↑	120	—	60	—	30	—

Note: * Not compatible with NMOS Specifications

Figure 2.13 Timing chart (Courtesy SGS)

2.8 PRACTICAL MICROCOMPUTER CIRCUITS

The circuit diagram of any small microcomputer is likely to contain some of the circuits associated with the address, data and control bus which have been discussed so far. For example, it will probably contain address bus buffers, address decoders and data bus buffers. In addition the clock circuit should be readily identifiable.

The Multitech Micro-Professor MPF-1B is a good example of a typical small computer system which contains a number of the features and chips that have been discussed (*Figures 2.14–2.16*, pages 44–6). It is used with the kind permission of Flight Electronics.

The Micro-Professor circuit is relatively straightforward, and it is therefore quite easy to identify the function of the various parts. The simplest way is to look at the circuits associated with each of the three buses in the system.

Data Bus

Starting with U1, the Z80 CPU, the data bus can be identified in the top right-hand corner. This bus is connected to each of the other main chips on sheet 1, U6, U7 and U8 as well as the main chips on sheets 2 and 3, U14, U11 and U10. Since there are only six devices connected to the data bus, there is no need for any buffering on the circuit board although if external devices are connected to the bus it would be worthwhile providing buffering for these additional loads. On sheet 2 the data bus is pulled up to +5 V with the RA_2 resistor pack, each value being 10K. Although this is not strictly necessary it maintains the data at FF hex whenever the data bus is in its tri-state condition. This helps to reduce any pick-up on the data bus lines and therefore maintains data integrity. In addition should any address be incorrectly used, such that no chip is found at that address, the data FF would be returned which will restore the operation to an address point within the monitor chip.

Address Bus

The address bus like the data bus has no special buffering simply because the number of chips on the circuit board is relatively small. The low-order address lines A_0–A_{11} go to the three main memory chips U6, U7 and U8. The high-order address lines A_{12}–A_{15} are used for address decoding. The address decoding chips are U5a and U5b together with U9a and U9b. 74LS139 devices are used which give an active low output. The main memory decoding is enabled by the MREQ signal from the Z80. Input/output decoding is also achieved with the 2–4 line decoder on Sheet 2. This decoding is sufficient for the 8255 I/O device U14, the counter timer U11 and the PIO U10. Note that the address inputs to this I/O decoder are A_6 and A_7 rather than address line A_{12} to A_{15}.

Control Bus

The control bus has also a number of very common features. For example, the clock circuit uses a standard 74LS14 inverter chip in an oscillator configuration to provide the basic clock waveform at 3.58 MHz. This is divided by two using a D-type flip-flop 74LS74 to generate the 1.79 MHz clock signal for the Z80. There is no special Z80 clock circuit, simply because the speed of operation at 1.79 MHz is relatively slow and therefore needs no special pull-up circuitry.

The main control signals are found at the top of the Z80 and again no buffering is provided. To the left of the Z80 CPU the INT, BUSREQ, and WAIT signals are each pulled to +5 V with a 10 K resistor. This is to ensure that they stay in their inactive state unless specifically required. The interrupt input can be enabled either with the INTR switch on the keyboard, shown on Sheet 1 or by an interrupt signal generated by one of the I/O devices U10 or U11.

The NMI signal is used as part of the single-step facility provided by the monitor program and is therefore connected via the 74LS90 to one of the pins of the 8255 I/O chip. In addition the Monitor key on the keyboard can be made to generate an Interrupt to this pin. The HALT signal is taken to Q3 which drives an LED, so that whenever the processor enters a halt state the light comes on.

The circuit diagrams of other common microcomputers, such as the BBC or IBM PC, are worth examining for their use of the circuits which have been discussed. In particular note how they use address buffers and decoders, together with

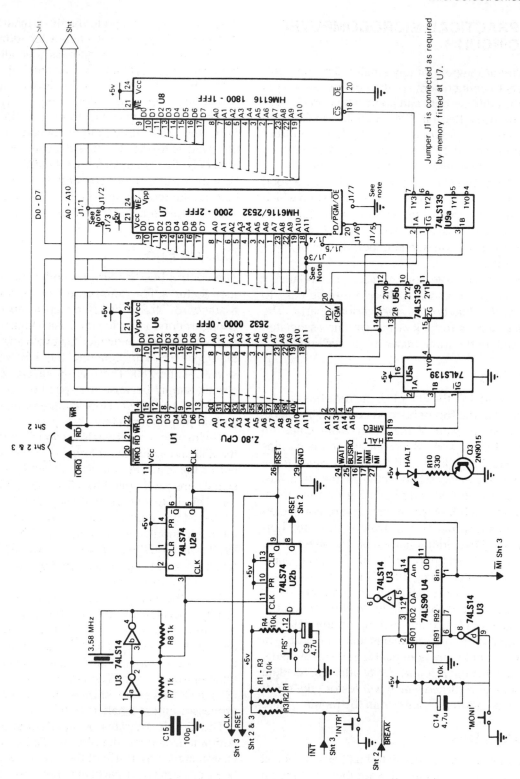

Figure 2.14   Micro-Professor sheet 1 (Courtesy Flight Electronics)

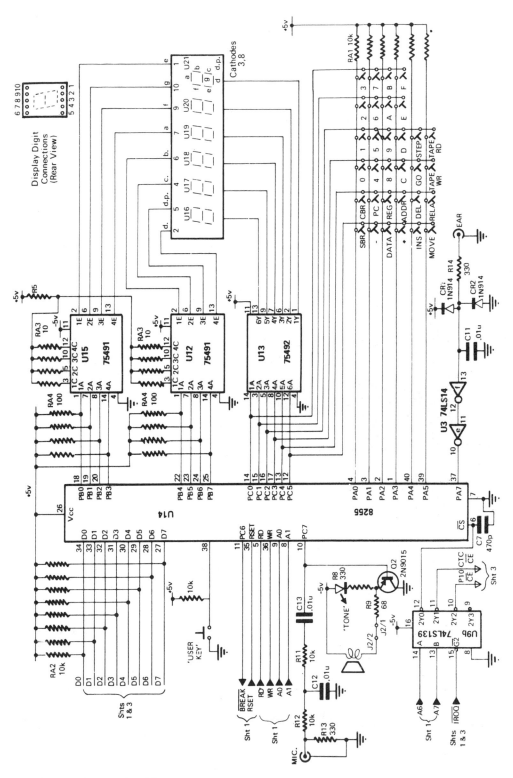

Figure 2.15 Micro-Professor sheet 2

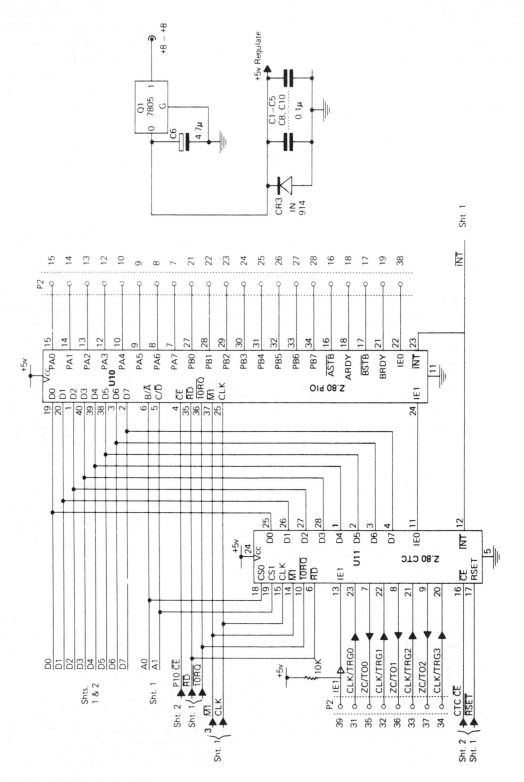

Figure 2.16   Micro-Professor sheet 3

buffers on the data bus. With different CPUs, the control signals will be rather different, but their basic function remains the same.

2.9 PRACTICAL BUS SIGNALS

When real systems are considered, the experimental results are not always exactly the same as those given in system data books. For example, the bus signals as seen on an oscilloscope are generally slightly different to the theoretical waveforms. Practical waveforms often have ringing, overshoots and undershoots and may not be very close to the ideal square waveforms which may be expected.

One way to examine real waveforms in a system is to force the processor into a loop by running a short program. This then creates repetitive waveforms which can be examined in detail.

Although an oscilloscope is not an ideal instrument for fault diagnosis in digital systems, it does provide true waveforms. Unfortunately, this is not always the case with other fault diagnosis tools such as logic analysers, which can provide idealised waveforms based on samples of the system logic levels.

To analyse the waveforms in a system it is first necessary to select a simple program. Anything but a very simple program would be too difficult to analyse using an oscilloscope and therefore only three instructions have been used. Although there are only three instructions, altogether ten machine cycles are required for each complete execution. The program is given below as a suitable example:

```
              ORG 1800H
    START:    LD A, (0555H)
              OUT (81H),A
              JP START
```

When this program is translated into machine code the hex values in *Table 2.1* are generated. These are shown with the appropriate memory addresses, and the appropriate machine cycle operation that takes place.

The bus activity in steps 4 and 7 relate to the execute part of the instructions in steps 1 and 5. Although these steps would not show by simple examination of the machine code program, they take place after each of these instructions and demand time on the system buses.

Table 2.1 Execution of program

Step	Address	Hex code	Cycle
1	1800	3A	Op-code fetch
2	1801	55	Memory read
3	1802	05	Memory read
4	0555	XX	Memory read
5	1803	D3	Op-code fetch
6	1804	81	Memory read
7	XX81	XX	Output write
8	1805	C3	Op-code fetch
9	1806	00	Memory read
10	1807	18	Memory read

To examine the waveforms in practice it is necessary to use an oscilloscope that is appropriately triggered so that each waveform can be related to each of the others. To do this channel 1 of the oscilloscope is connected to the $\overline{WR}$ pin of the microprocessor, which is used as the trigger for the oscilloscope and provides a means of obtaining a stable trace from the system. Channel 2 of the oscilloscope is used to display each of the subsequent waveforms as required in the correct time relationship with the trigger channel.

Typical system waveforms are shown in *Figure 2.17* (overleaf). They clearly illustrate a number of important features of practical waveforms. The logic levels are not constant. For example, the $\overline{WR}$ signal exhibits significant noise at its logic 1 level. The address lines A_0–A_6 have a number of pulses, shown dotted, which are the 'refresh addresses' which appear on the bus. This allows the negative-going $\overline{MREQ}$ pulses, numbers 2, 6 and 11 to be identified as the refresh parts of the cycles.

The data bus signals appear to 'float' towards a logic 1 level when they are not active, which gives rise to small triangular pulses on the bus lines.

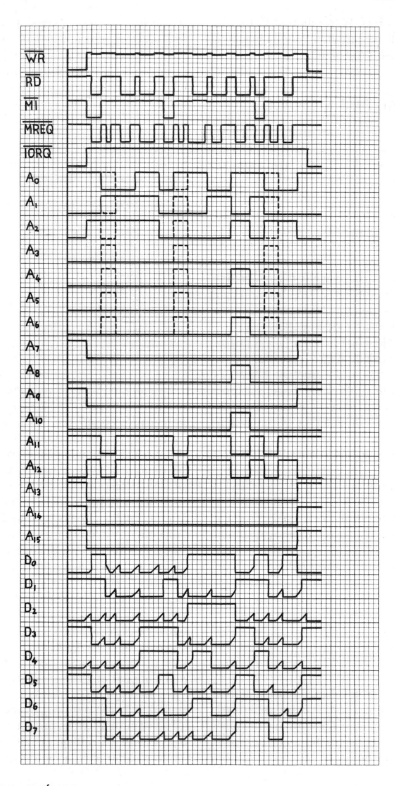

Figure 2.17 System waveforms

Summary

The main points covered in this chapter relate to the operation of the data and control buses and the types of circuit associated with each:

- Control bus signals vary between microprocessors but they all control how a system operates at any moment.

- Tri-state devices must be used on the data bus.

- Bidirectional bus buffers or bus transceivers provide data bus buffering in both directions.

- Real signals in microcomputer systems are much less than 'ideal' when compared with those in data sheets.

Questions

2.1 Why is it necessary to ensure that only one device connected to the data bus can output data at any moment?

2.2 What characteristic must all devices connected to the data bus have in common?

2.3 What is the function of a 'chip enable' signal?

2.4 Which control signals need to be used to correctly enable an EPROM?

2.5 Why is it not necessary for all systems to include a bidirectional buffer?

2.6 List the main control signals in the CPU control group for a Z80 processor.

2.7 Why does a Z80 require a special clock circuit?

2.8 What is the function of the $\overline{\text{BUSREQ}}$ and $\overline{\text{BUSACK}}$ signals?

2.9 Explain the sentence 'The microprocessor is the only "talker" on the address bus'.

2.10 Why are tri-state devices required for connection to the data bus in a system?

2.11 What characteristics must be true of a system with no bus drivers?

THREE

Processors and input/output devices

When you have completed this chapter you should be able to:

1 Identify, using manufacturer's literature, the characteristics of a single-chip computing element, such as an 8-bit, 16-bit, or a bit-slice microprocessor.

2 Discuss, using manufacturer's literature, the function, operation and distinguishing characteristics of:
(a) Parallel input and output (I/O) ports.
(b) Bidirectional I/O port.
(c) Programmable I/O ports.
(d) Parallel-to-serial output port.
(e) Programmable serial I/O ports.

3.1 MICROPROCESSOR CHARACTERISTICS

The microcomputer chip is invariably the most complex VLSI device in a computer circuit. Some modern peripheral control devices, such as floppy disc controller chips, are almost as complex but generally the microprocessor itself is more complicated. This means that its data sheets are also highly complicated, and to specify every possible aspect of the microprocessor's functions, the data sheets are more correctly described as data books.

To a system designer the Microcomputer Data Book is as vital as a map to a foreign traveller. Without the data on an integrated circuit it is almost impossible to design a system that will function correctly.

Manufacturer's literature must contain every conceivable piece of information that a system designer or user needs to know to use the chip to its best advantage. It is also necessary to have all the information required to ensure that the device is always operated within its specification, so that its operation will always be predictable and repeatable.

Manufacturers must ensure that the data sheets contain information on every possible use of the device, and give information on its performance in its full range of operating conditions. Once the specification is set, then it is their responsibility to ensure that all production devices meet the specification or exceed it.

Most microprocessor data sheets are structured in a similar way, so that once the information from one has been read, it is relatively easy to find similar information on the data sheets of other microprocessors. However, all manufacturers are at liberty to produce data sheets to their own specification so that not every data sheet will contain information in precisely the same format. A typical data sheet would contain the following type of information.

Introduction and main features summary

The first page of most data sheets contains a summary of the main features of the microprocessor. For example, this will include information on the type of technology used, the size of the data and address buses and the main power supply requirements. It may also refer to any special features of the microprocessor that make it a unique device. This may include a number of instructions, or its speed, or its manufacturing technology.

Pin assignment diagram

The pin assignment diagram is a vital piece of information for every designer or microprocessor user at competent level. It indicates the function of each pin on the device. Without it, incorporation of a microprocessor into a circuit would be impossible.

Programmer's register model

All microprocessors contain a series of registers and a programmer must know the names of these together with their functions or special features. Most software designers would need this information before any software could be produced at assembly language level. This diagram may also be supplemented by an internal block diagram of the microprocessor itself, if this contains other features of particular interest.

Absolute maximum ratings

Most data sheets contain a table known as the **absolute maximum ratings table**. This gives the maximum permitted power supply voltage that the device can take before it is destroyed. It may also indicate the maximum possible operating temperature range. Any voltage or temperature outside the specification range is likely to damage the device permanently.

d.c. electrical characteristics

d.c. electrical characteristics indicate the normal operating conditions for the device. Typically these would include the supply voltages, the maximum/minimum logic levels, the input and output leakage currents, etc. They give the circuit

designer an idea of the load that the component will place on the circuit when it is operated.

Test conditions circuit

Most data sheets also contain information on the circuits used to test the component. This allows the designer to determine whether the circuit conditions to be applied are equivalent to those under which the device was tested.

a.c. electrical specification

The a.c. electrical specification for a microprocessor is generally a very large chart which contains details of the timing values related to the circuit waveforms. Every waveform from every pin is given a maximum and minimum timing value relative to a fixed point in the instruction or machine cycle. This allows designers to predict precisely when logic levels will change. Timing values relate to the main read, write and input/output cycles which the microprocessor operates.

Bus cycle descriptions

Most microprocessor data sheets contain detailed descriptions of each bus cycle, specifically referring to the bus cycle timing chart. For example, the operation of the interrupt line and the actions taken by the microprocessor would need to be explained in detail in this section of the data sheet, because most microprocessors differ in this respect.

Instruction set and instruction timing

The microprocessor not only requires a physical description, but the range of instructions that it can execute also need detailed description. Some manufacturers choose to put their instruction set and timing information in separate data books, although it is possible to have them all included with the electrical characteristics. Not only is the description of the actual instruction required, but also a detailed description of the number of T-states and machine cycles required. This allows programmers to determine precisely what will happen when specific instructions are executed, and the time the program will take.

Physical package dimensions

The last section of most data sheets contains a physical diagram of the component and its dimensions. This is specially for the board designer who needs to produce a printed circuit board that will accept the component. In addition it is important information for the designers of sockets and automatic insertion machines.

The information supplied for an individual microprocessor must be extensive. Typically it may run for 40 or 50 pages. Therefore it would be impossible to reproduce all of that information in this book, and it is suggested that if possible, an actual microcomputer data book is obtained and the complete data sheets of a microprocessor investigated.

However, some aspects of such a data sheet have been extracted and the following case studies to illustrate one or two of the typical features.

3.2 CASE STUDY 1 – THE Z80 MICROPROCESSOR

Since the Z80 has been the main microprocessor studied so far, it is appropriate to consider extracts from its data book first. Just one page has been extracted from the extensive Z80 data sheet for further examination.

The first page (*Figure 3.1*) contains a description of the main features of the Z80 together with a diagram showing its main logic functions. None of this information will be new, but it is interesting to know that the manufacturers place the high priority on this type of information by putting it as the first page of the data sheet.

The sheet continues with a general description of the Z80 and a detailed examination of the instruction set, all of which will be very familiar. After the instruction set details the data sheet continues with timing diagrams, again which will

not be new. Each timing diagram is associated with a series of times precisely specified, and tabulated in the form of the a.c. characteristics of the device.

This is followed by the Z80 d.c. characteristics. Although these are not remarkable in any way they do indicate that the Z80 is designed to be totally TTL compatible in its logic levels, as well as indicating that the typical power supply requirements vary between 150 and 200 mA under normal operating conditions.

It also shows that the input capacitance of the clock circuit at 35 pF is more than three times the output capacitance and seven times the normal input capacitance of other pins. This is one of the reasons that the Z80 manufacturers advise of the use of a special clock circuit which was discussed earlier.

3.3 CASE STUDY 2 – THE 68000 MICROPROCESSOR

The 68000 is a 16-bit microprocessor, of a later generation than the Z80. It is very different in many respects.

Just four pages have been extracted from its very extensive data sheet, to give an indication of the main characteristics of the device (*Figure 3.2*, pages 54–7).

The most obvious difference between this and a Z80 is that the 68000 is a 64-pin device, with 23 address lines and 16 data lines. The 16 data lines mask the fact that internally the registers are 32 bits wide and this allows great flexibility in its functions. Its directly addressable memory is 16 Mbytes.

As usual the main features of the microprocessor are highlighted on the first page of its data sheet, together with the programmer's model and pin assignment for the chip.

Apart from the maximum ratings, and the main electrical characteristics of the 68000, the diagrams show that the timing information applies to a range of devices that vary in clock frequency from 4 MHz up to 12.5 MHz.

The **read** cycle timing diagram is included to give a brief insight into the complexity of the operation of this 16 pin device. In comparison to

Z80 CPU Central Process Unit

- The instruction set contains 158 instructions. The 78 instructions of the 8080A are included as a subset; 8080A and Z80* software compatibility is maintained.

- 8MHz, 6MHz, 4MHz and 2.5 MHz clocks for the Z80H, Z80B, Z80A and Z80 CPU result in rapid instruction execution with consequent high data throughput.

- The extensive instruction set includes string, bit, byte, and word operations. Block searches and block transfers together with indexed and relative addressing result in the most powerful data handling capabilities in the microcomputer industry.

- The Z80 microprocessors and associated family of peripheral controllers are linked by a vectored interrupt system. This system may be daisy-chained to allow implementation of a priority interrupt scheme. Little, if any, additional logic is required for daisy-chaining.

- Duplicate sets of both general-purpose and flag registers are provided, easing the design and operation of system software through single-context switching, background-foreground programming, and single-level interrupt processing. In addition, two 16-bit index registers facilitate program processing of tables and arrays.

- There are three modes of high speed interrupt processing: 8080 similar, non-Z80 peripheral device, and Z80 Family peripheral with or without daisy chain.

- On-chip dynamic memory refresh counter.

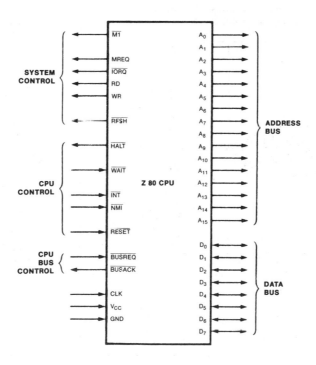

Figure 3.1   Z80 data sheet extracts (Courtesy SGS)

16-BIT MICROPROCESSING UNIT

Advances in semiconductor technology have provided the capability to place on a single silicon chip a microprocessor at least an order of magnitude higher in performance and circuit complexity than has been previously available. The MC68000 is the first of a family of such VLSI microprocessors from Motorola. It combines state-of-the-art technology and advanced circuit design techniques with computer sciences to achieve an architecturally advanced 16-bit microprocessor.

The resources available to the MC68000 user consist of the following:

- 32-Bit Data and Address Registers
- 16 Megabyte Direct Addressing Range
- 56 Powerful Instruction Types
- Operations on Five Main Data Types
- Memory Mapped I/O
- 14 Addressing Modes

As shown in the programming model, the MC68000 offers seventeen 32-bit registers in addition to the 32-bit program counter and a 16-bit status register. The first eight registers (D0-D7) are used as data registers for byte (8-bit), word (16-bit), and long word (32-bit) data operations. The second set of seven registers (A0-A6) and the system stack pointer may be used as software stack pointers and base address registers. In addition, these registers may be used for word and long word address operations. All seventeen registers may be used as index registers.

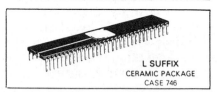

MC68000L4 (4 MHz) **MC68000L6** (6 MHz)

MC68000L8 (8 MHz) **MC68000L10** (10 MHz)

MC68000L12 (12.5 MHz)

HMOS
(HIGH-DENSITY, N-CHANNEL, SILICON-GATE DEPLETION LOAD)

16-BIT MICROPROCESSOR

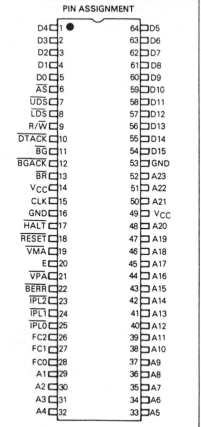

L SUFFIX
CERAMIC PACKAGE
CASE 746

PIN ASSIGNMENT

D4	1 ●	64 D5
D3	2	63 D6
D2	3	62 D7
D1	4	61 D8
D0	5	60 D9
$\overline{AS}$	6	59 D10
$\overline{UDS}$	7	58 D11
$\overline{LDS}$	8	57 D12
$R/\overline{W}$	9	56 D13
$\overline{DTACK}$	10	55 D14
$\overline{BG}$	11	54 D15
$\overline{BGACK}$	12	53 GND
$\overline{BR}$	13	52 A23
V_{CC}	14	51 A22
CLK	15	50 A21
GND	16	49 V_{CC}
$\overline{HALT}$	17	48 A20
$\overline{RESET}$	18	47 A19
$\overline{VMA}$	19	46 A18
E	20	45 A17
$\overline{VPA}$	21	44 A16
$\overline{BERR}$	22	43 A15
$\overline{IPL2}$	23	42 A14
$\overline{IPL1}$	24	41 A13
$\overline{IPL0}$	25	40 A12
FC2	26	39 A11
FC1	27	38 A10
FC0	28	37 A9
A1	29	36 A8
A2	30	35 A7
A3	31	34 A6
A4	32	33 A5

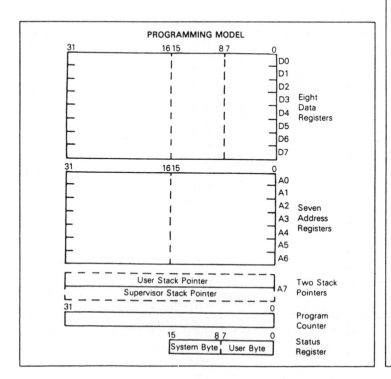

PROGRAMMING MODEL

Eight Data Registers (D0–D7)

Seven Address Registers (A0–A6)

User Stack Pointer / Supervisor Stack Pointer — A7 — Two Stack Pointers

Program Counter

Status Register — System Byte | User Byte

Figure 3.2(a)–(d) The 68000 data sheet extracts (Courtesy Motorola)

MAXIMUM RATINGS

Rating	Symbol	Value	Unit
Supply Voltage	V_{CC}	− 0.3 to + 7.0	V
Input Voltage	V_{in}	− 0.3 to + 7.0	V
Operating Temperature Range	T_A	0 to 70	°C
Storage Temperature	T_{stg}	− 55 to 150	°C

THERMAL CHARACTERISTICS

Characteristic	Symbol	Value	Unit
Thermal Resistance Ceramic Package	θ_{JA}	30	°C/W

This device contains circuitry to protect the inputs against damage due to high static voltages or electric fields; however, it is advised that normal precautions be taken to avoid application of any voltage higher than maximum-rated voltages to this high-impedance circuit. Reliability of operation is enhanced if unused inputs are tied to an appropriate logic voltage level (e.g., either V_{SS} or V_{CC}.

POWER CONSIDERATIONS

The average chip-junction temperature, T_J, in °C can be obtained from:

$$T_J = T_A + (P_D \cdot \theta_{JA}) \tag{1}$$

Where:

$T_A \equiv$ Ambient Temperature, °C

$\theta_{JA} \equiv$ Package Thermal Resistance, Junction-to-Ambient, °C/W

$P_D \equiv P_{INT} + P_{I/O}$

$P_{INT} \equiv I_{CC} \times V_{CC}$, Watts − Chip Internal Power

$P_{I/O} \equiv$ Power Dissipation on Input and Output Pins − User Determined

For most applications $P_{I/O} \blacktriangleleft P_{INT}$ and can be neglected.

An approximate relationship between P_D and T_J (if $P_{I/O}$ is neglected) is:

$$P_D = K \div (T_J + 273°C) \tag{2}$$

Solving equations 1 and 2 for K gives:

$$K = P_D \cdot (T_A + 273°C) + \theta_{JA} \cdot P_D^2 \tag{3}$$

Where K is a constant pertaining to the particular part. K can be determined from equation 3 by measuring P_D (at equilibrium) for a known T_A. Using this value of K the values of P_D and T_J can be obtained by solving equations (1) and (2) iteratively for any value of T_A.

DC ELECTRICAL CHARACTERISTICS (V_{CC} = 5.0 Vdc ± 5%, V_{SS} = 0 Vdc; T_A = 0°C to 70°C. See Figures 1, 2, and 3)

Characteristic		Symbol	Min	Max	Unit
Input High Voltage		V_{IH}	2.0	V_{CC}	V
Input Low Voltage		V_{IL}	V_{SS} − 0.3	0.8	V
Input Leakage Current @ 5.25 V	$\overline{BERR}$, $\overline{BGACK}$, $\overline{BR}$, $\overline{DTACK}$, CLK, $\overline{IPL0-IPL2}$, $\overline{VPA}$	I_{in}	−	2.5	μA
	$\overline{HALT}$, $\overline{RESET}$		−	20	
Three-State (Off State) Input Current @ 2.4 V/0.4 V	$\overline{AS}$, A1-A23, D0-D15 FC0-FC2, $\overline{LDS}$, R/$\overline{W}$, $\overline{UDS}$, $\overline{VMA}$	I_{TSI}	−	20	μA
Output High Voltage (I_{OH} = − 400 μA)	E*	V_{OH}	V_{CC} − 0.75	−	V
	$\overline{AS}$, A1-A23, $\overline{BG}$, D0-D15 FC0-FC2, $\overline{LDS}$, R/$\overline{W}$, $\overline{UDS}$, $\overline{VMA}$		2.4	−	
Output Low Voltage (I_{OL} = 1.6 mA) (I_{OL} = 3.2 mA) (I_{OL} = 35.0 mA) (I_{OL} = 5.3 mA)	$\overline{HALT}$ A1-A23, $\overline{BG}$, FC0-FC2 $\overline{RESET}$ E, $\overline{AS}$, D0-D15, $\overline{LDS}$, R/$\overline{W}$ $\overline{UDS}$, $\overline{VMA}$	V_{OL}	− − − −	0.5 0.5 0.5 0.5	V
Power Dissipation (Clock Frequency = 8 MHz)		P_D	−	1.5	W
Capacitance (V_{in} = 0 V, T_A = 25°C; Frequency = 1 MHz)		C_{in}	−	10.0	pF

* With external pullup resistor of 470 Ω

Figure 3.2(b)

FIGURE 1 — RESET TEST LOAD

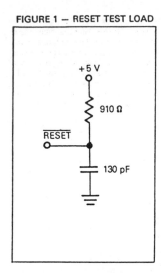

FIGURE 2 — HALT TEST LOAD

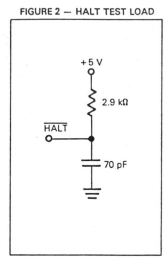

FIGURE 3 — TEST LOADS

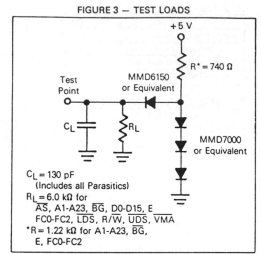

CLOCK TIMING (See Figure 4)

Characteristic	Symbol	4 MHz MC68000L4		6 MHz MC68000L6		8 MHz MC68000L8		10 MHz MC68000L10		12.5 MHz MC68000L12		Unit
		Min	Max	Min	Max	Min	Max	Min	Max	Min	Max	
Frequency of Operation	F	2.0	4.0	2.0	6.0	2.0	8.0	2.0	10.0	4.0	12.5	MHz
Cycle Time	t_{cyc}	250	500	167	500	125	500	100	500	80	250	ns
Clock Pulse Width	t_{CL}	115	250	75	250	55	250	45	250	35	125	ns
	t_{CH}	115	250	75	250	55	250	45	250	35	125	
Rise and Fall Times	t_{Cr}	—	10	—	10	—	10	—	10	—	5	ns
	t_{Cf}	—	10	—	10	—	10	—	10	—	5	

FIGURE 4 — INPUT CLOCK WAVEFORM

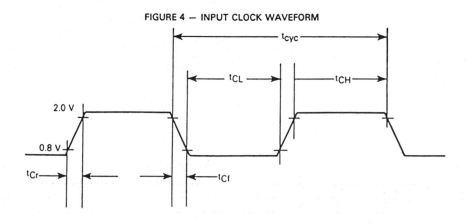

Figure 3.2(c)

FIGURE 5 — READ CYCLE TIMING

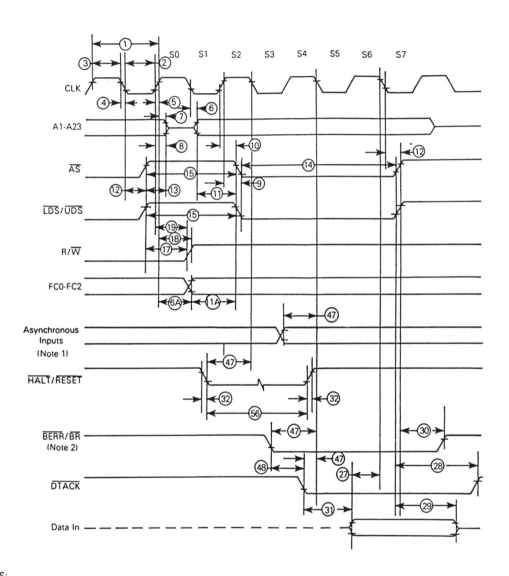

NOTES:

1. Setup time for the asynchronous inputs $\overline{\text{BGACK}}$, $\overline{\text{IPL0-IPL2}}$, and $\overline{\text{VPA}}$ guarantees their recognition at the next falling edge of the clock.
2. $\overline{\text{BR}}$ need fall at this time only in order to insure being recognized at the end of this bus cycle.
3. Timing measurements are referenced to and from a low voltage of 0.8 volts and a high voltage of 2.0 volts, unless otherwise noted.

Figure 3.2(d)

the Z80, it is clear that there are a number of additional control lines which may need to be taken into consideration, and which can sometimes affect the operation of the **read** cycle.

Unlike the Z80, the 68000 is an **asynchronous** device, and does not operate on a fixed time per cycle. Instead it waits for an acknowledge signal to be returned from the memory or an I/O device to tell the processor that the information is ready and is valid. This is the DTACK signal which indicates that the data is available on the bus. When the microprocessor receives the signal, information is read in during the next clock cycle.

3.4 BIT-SLICE PROCESSORS

Bit-slice processors are not microprocessors in their own right, but are actually slices of microprocessors. As such they contain a part of arith-metic logic unit, together with its multiplexers and data paths. In addition, it may also include registers, flags and buses. Typically a bit-slice processor will contain four bits, so that ALUs of different sizes may be built up simply by adding together the requisite number of bits. Most bit-slice processors need extra chips to build the control unit and its microprogram function, so that a complete bit-slice processor may contain six or seven individual chips.

The real purpose in having a bit-slice processor is to enable high-speed computers to be designed and built, of different size, so that system designers are not restricted to the range of microprocessors currently available. This allows mini-computers of varying complexity to be easily constructed. Most semiconductor manufacturers produce or have produced bit slice devices, and these include National Semiconductor, Intel, Motorola, and Texas Instruments.

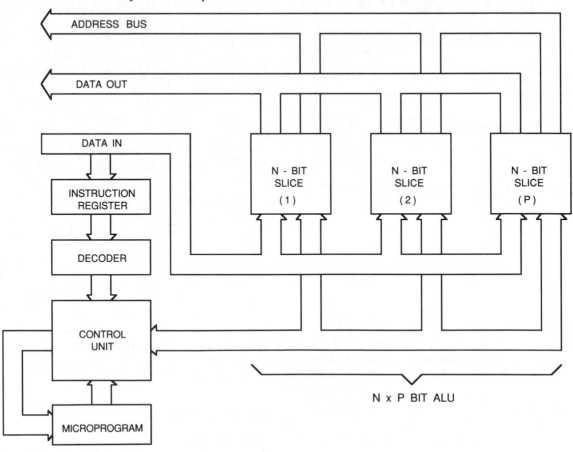

Figure 3.3   Bit slice processor

Probably the most well known bit-slice processor was the 2901, from Advanced Micro Devices. This was a 4-bit bit slice, with high speed, typically 105 ns, and its architecture is shown in *Figure 3.3*. Apart from the fact that large ALUs can be built using sections of the device, it had additional capabilities such as improved arithmetic processing with the addition of a Carry-Look Ahead Unit in discrete components, and additional control logic, loop counters and bus multiplexing capability. Bit-slice processors are widely used in micro- and mini-computer designs, as well as in many military processors.

3.5 INPUT/OUTPUT PORTS

All microcomputer systems need to input data from other systems and output results. There are various circuits that allow microcomputers to do this and in general they are known as either **input** or **output** ports. In the most simple systems an input port may simply be an octal buffer, such as a 74LS244, which provides 8 input bits when requested by the CPU. Similarly, to obtain output from a microprocessor system, all that is required is a simple latch, such as a 74LS373.

These devices are fine for basic operations, but where more complex functions are required such as interrupt facilities, or bidirectional operation, then a more comprehensive I/O chip is required. Most systems would use a programmable I/O device for such purposes since these can be programmed to act in any way the programmer desires and their function can be changed within the software if necessary. This simplifies the design of the hardware by shifting some of the burden onto the software in the system.

Simple Input Port – The 74LS244

In the manufacturer's data sheets, the 74LS244 is described as an octal buffer and line driver with three state outputs. As such it is one of a series of devices with similar characteristics with slightly different functions. Its data sheet is shown in a previous chapter (pages 10–12).

From the data sheet it can be seen that the 74LS244 is actually two groups of four buffers

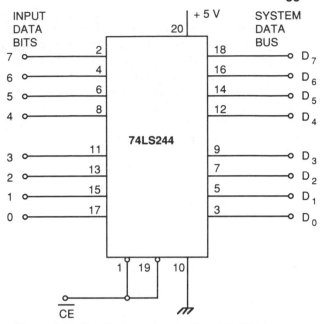

Figure 3.4   Practical input port with a 74LS244

each controlled by a single enable pin. The enable pin is active low so that the devices are turned on with a logic '0'. When either enable pin is in its logic '1' state then four of the buffers are in their tri-state condition. Each buffer is capable of driving a relatively large current and this is typically 24 mA at a logic '0' level and 15 mA at a logic '1' level.

To connect the device as an octal buffer, pins 1 and 19 will be connected together and enabled by a suitable chip enable signal, derived in a Z80 based system from $\overline{IORQ}$, $\overline{RD}$ and a suitable port address. A typical circuit is shown in *Figure 3.4*.

The main advantage of such a circuit is that it is very simple, and therefore easy to implement. No programming is required other than executing an input instruction using the port address chosen by a design of the hardware. The main disadvantage of such a system is that there is no facility for the port to generate an interrupt except by adding an additional logic circuit to the input data bits, and in some circumstances this may prove to be relatively complex.

Simple Output Port

In the same way that a tri-state buffer can be used as an input port, an octal latch can be used as a

TYPES SN54LS373, SN54LS374, SN54S373, SN54S374, SN74LS373, SN74LS374, SN74S373, SN74S374 OCTAL D-TYPE TRANSPARENT LATCHES AND EDGE-TRIGGERED FLIP-FLOPS

BULLETIN NO. DL-S 7612350, OCTOBER 1976

TTL
MSI

- Choice of 8 Latches or 8 D-Type Flip-Flops In a Single Package
- 3-State Bus-Driving Outputs
- Full Parallel-Access for Loading
- Buffered Control Inputs
- Clock/Enable Input Has Hysteresis to Improve Noise Rejection
- P-N-P Inputs Reduce D-C Loading on Data Lines ('S373 and 'S374)

SN54LS373, SN54S373 . . . J PACKAGE
SN74LS373, SN74S373 . . . J OR N PACKAGE
(TOP VIEW)

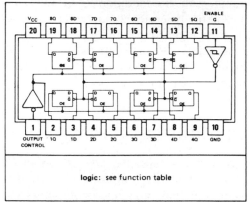

logic: see function table

SN54LS374, SN54S374 . . . J PACKAGE
SN74LS374, SN74S374 . . . J OR N PACKAGE
(TOP VIEW)

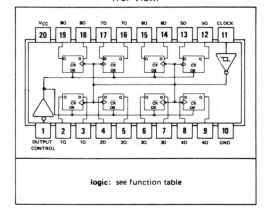

logic: see function table

'LS373, 'S373
FUNCTION TABLE

OUTPUT CONTROL	ENABLE G	D	OUTPUT
L	H	H	H
L	H	L	L
L	L	X	Q_0
H	X	X	Z

'LS374, 'S374
FUNCTION TABLE

OUTPUT CONTROL	CLOCK	D	OUTPUT
L	↑	H	H
L	↑	L	L
L	L	X	Q_0
H	X	X	Z

See explanation of function tables on page 3-8.

description

These 8-bit registers feature totem-pole three-state outputs designed specifically for driving highly-capacitive or relatively low-impedance loads. The high-impedance third state and increased high-logic-level drive provide these registers with the capability of being connected directly to and driving the bus lines in a bus-organized system without need for interface or pull-up components. They are particularly attractive for implementing buffer registers, I/O ports, bidirectional bus drivers, and working registers.

The eight latches of the 'LS373 and 'S373 are transparent D-type latches meaning that while the enable (G) is high the Q outputs will follow the data (D) inputs. When the enable is taken low the output will be latched at the level of the data that was setup.

Figure 3.5(a)–(b)   74LS373 and 74LS374 data sheets (Courtesy Texas Instruments)

TYPES SN54LS373, SN74LS374, SN54S373, SN54S374, SN74LS373, SN74LS374, SN74S373, SN74S374
OCTAL D-TYPE TRANSPARENT LATCHES AND
EDGE-TRIGGERED FLIP-FLOPS

description (continued)

The eight flip-flops of the 'LS374 and 'S374 are edge-triggered D-type flip-flops. On the positive transition of the clock, the Q outputs will be set to the logic states that were setup at the D inputs.

Schmitt-trigger buffered inputs at the enable/clock lines simplify system design as ac and dc noise rejection is improved by typically 400 mV due to the input hysteresis. A buffered output control input can be used to place the eight outputs in either a normal logic state (high or low logic levels) or a high-impedance state. In the high-impedance state the outputs neither load nor drive the bus lines significantly.

The output control does not affect the internal operation of the latches or flip-flops. That is, the old data can be retained or new data can be entered even while the outputs are off.

logic symbol

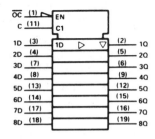

logic diagram (positive logic)

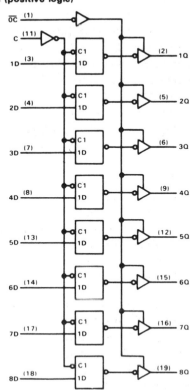

logic symbol

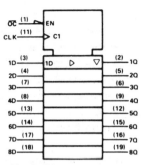

logic diagram (positive logic)

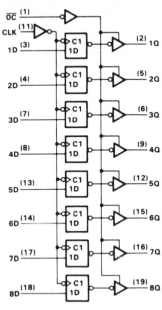

Figure 3.5(b)

simple output port. A typical device from the 74 series is the 74LS373, which provides 8 'D' type latches with a single enable pin, and a tri-state output capability (*Figure 3.5*, see pages 60–1). Under normal circumstances the tri-state output would not be required and therefore it could be permanently enabled to provide either a logic '1' or a logic '0' output.

From the data sheets it can be seen that the LS373 and LS374 devices are very similar. The main difference is in the triggering of the flip-flops since the LS374 is an edge-triggered device and operates on the rising edge of the clock pulse whereas the LS373 is a level-triggered device and only operates when the clock is in its high logic state. The edge triggered device may be useful in applications where the data is likely to change during the time the clock is in its high logic state.

These octal latches are also capable of driving a reasonably high current, and each one can be used as a bus driver. The enable input contains a Schmitt-trigger and this reduces any system noise and improves reliability in the operation of the device.

A typical output port using a 74LS373 is shown in *Figure 3.6*. As with the input port the chip enable is derived from a combination of $\overline{IORQ}$, $\overline{WR}$ and a suitable memory or I/O address.

The simple output port provides eight data bits but is restricted if modifications to the port are

required. For example, if a system required seven output bits and one input bit, it would be necessary to have two 8 bit chips in the system. Also, a simple port like this has no mechanism for synchronising data transfers between the port and the peripheral devices to which it is connected. This also requires additional logic circuits.

Programmable Input/Output

Programmable input/output devices are a particularly important form of interfacing circuit with wide application in microelectronic systems. Their main advantages are that within a 40-pin device, at least two I/O ports, all the control circuits, handshaking logic and interrupt control logic can be contained. This reduces the number of components required in the system considerably. In addition, they have the added flexibility that they are programmable so that they may be adapted for different requirements under software control. Different manufacturers call their programmable I/O devices by different names, such as **PIC (peripheral interface circuit)** or **PPI (programmable peripheral interface)** or **PIO (programmable input/output)**, which all refer to the same type of device.

Most programmable I/O devices have a number of features in common:

(a) They contain two or three data ports.
(b) They contain control ports which determine the operation of the data ports.
(c) They contain interface circuits which allow communication between the ports and the peripheral devices.
(d) Ports can be operated as inputs, outputs or in bidirectional mode.
(e) Ports may be allowed to generate interrupts to the processor.

The block diagram of an Intel 8255 I/O device is shown in *Figure 3.7*. This is a relatively straightforward device with three 8-bit ports and a control register which occupy four separate port addresses. The contents of the control register determine the function of the three data ports.

The device is programmed by writing 8 bits of data into the control register. Each bit has a

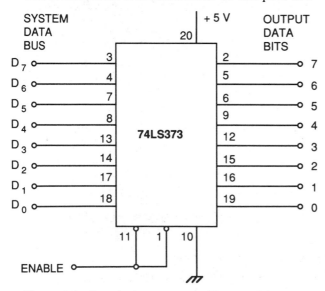

Figure 3.6 Practical output port with a 74LS373

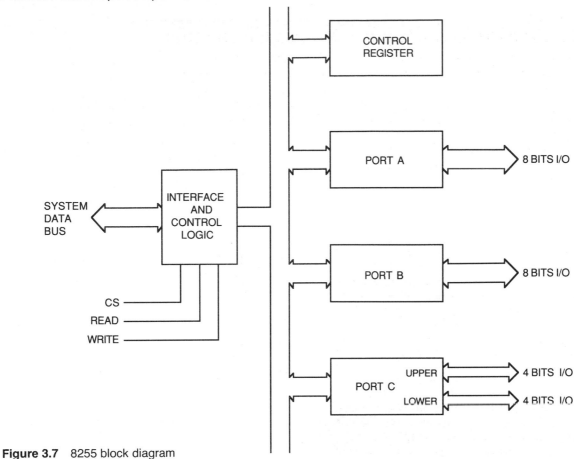

Figure 3.7 8255 block diagram

specific function and is used to control the flow of data in one of the data ports. A full description of this device would take too much space but, for example, it is possible to use ports B and C as outputs while port A is configured as an input. This requires a control byte of 90 hex to be sent to the control port. *Figure 3.8* shows how each bit affects the data ports. The data is written into the

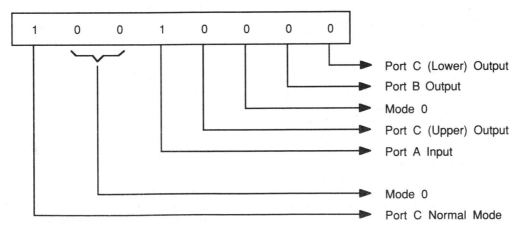

Figure 3.8 8255 initialisation

Z80 PIO Parallel Input/Output Controller

- Provides a direct interface between Z80 microcomputer systems and peripheral devices.

- Both ports have interrupt-driven handshake for fast response.

- Four programmable operating modes: byte input, byte output, byte input/output (Port A only), and bit input/output.

- Programmable interrupts on peripheral status conditions.

- Standard Z80 Family bus-request and prioritized interrupt-request daisy chains implemented without external logic.

- The eight Port B outputs can drive Darlington transistors (1.5 mA at 1.5 V).

General Description

The Z80 PIO Parallel I/O Circuit is a programmable, dual-port device that provides a TTL-compatible interface between peripheral devices and the Z80 CPU. The CPU configures the Z80 PIO to interface with a wide range of peripheral devices with no other external logic. Typical peripheral devices that are compatible with the Z80 PIO include most keyboards, paper tape readers and punches, printers, PROM programmers, etc.

One characteristic of the Z80 peripheral controllers that separates them from other interface controllers is that all data transfer between the peripheral device and the CPU is accomplished under interrupt control. Thus, the interrupt logic of the PIO permits full use of the efficient interrupt capabilities of the Z80 CPU during I/O transfers. All logic necessary to implement a fully nested interrupt structure is included in the PIO.

Another feature of the PIO is the ability to interrupt the CPU upon occurrence of specified status conditions in the peripheral device. For example, the PIO can be programmed to interrupt if any specified peripheral alarm conditions should occur. This interrupt capability reduces the time the processor must spend in polling peripheral status.

The Z80 PIO interfaces to peripherals via two independent general-purpose I/O ports, designated Port A and Port B. Each port has eight data bits and two handshake signals, Ready and Strobe, which control data transfer. The Ready output indicates to the peripheral that the port is ready for a data transfer. Strobe is an input from the

peripheral that indicates when a data transfer has occured.

Operating Modes. The Z80 PIO ports can be programmed to operate in four modes: byte output (Mode 0), byte input (Mode 1), byte input/output (Mode 2) and bit input/output (Mode 3).

In Mode 0, either Port A or Port B can be programmed to output data. Both ports have output registers that are individually addressed by the CPU; data can be written

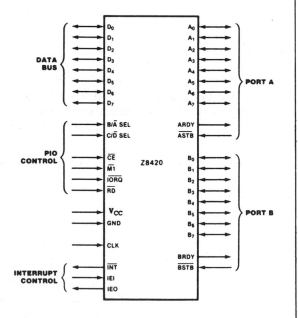

Figure 1. Logic Functions

Figure 3.9(a)–(d)   Z80 PIO data sheets (Courtesy SGS)

General Description (Continued)

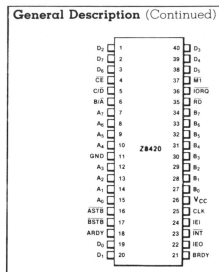

Figure 2. Pin Configuration

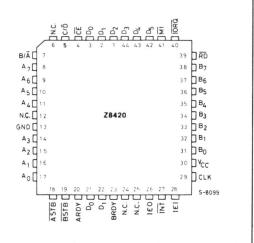

Figure 2a. Chip Carrier Pin Configuration

to either port at any time. When data is written to a port, an active Ready output indicates to the external device that data is available at the associated port and is ready for transfer to the external device. After the data transfer, the external device responds with an active Strobe input, which generates an interrupt, if enabled.

In Mode 1, either Port A or Port B can be configured in the input mode. Each port has an input register addressed by the CPU. When the CPU reads data from a port, the PIO sets the Ready signal, which is detected by the external device. The external device then places data on the I/O lines and strobes the I/O port, which latches the data into the Port Input Register, resets Ready, and triggers the Interrupt Request, if enabled. The CPU can read the input data at any time, which again sets Ready.

Mode 2 is bidirectional and uses Port A, plus the interrupts and handshake signals from both ports. Port B must be set to Mode 3 and masked off. In operation, Port A is used for both data input and output. Output operation is similar to Mode 0 except that data is allowed out onto the Port A bus only when $\overline{ASTB}$ is Low. For input, operation is similar to Mode 1, except that the data input uses the Port B handshake signals and the Port B interrupt (if enabled).

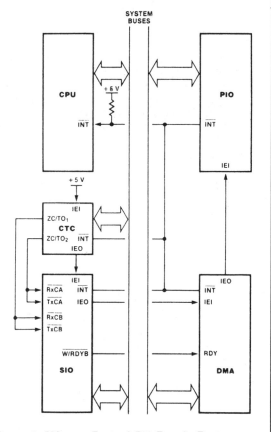

Figure 3. PIO in a Typical Z80 Family Environment

Figure 3.9(b)

General Description (Continued)

Both ports can be used in Mode 3. In this mode, the individual bits are defined as either input or output bits. This provides up to eight separate, individually defined bits for each port. During operation, Ready and Strobe are not used. Instead, an interrupt is generated if the condition of one input changes, or if all inputs change. The requirements for generating an interrupt are defined during the programming operation; the active level is specified as either High or Low, and the logic condition is specified as either one input active (OR) or all inputs active (AND). For example, if the port is programmed for active Low inputs and the logic function is AND, then all inputs at the specified port must go Low to generate an interrupt.

Data outputs are controlled by the CPU and can be written or changed at any time.

- Individual bits can be masked off.
- The handshake signals are not used in Mode 3, Ready is held Low, and Strobe is disabled.
- When using the Z80 PIO interrupts, the Z80 CPU interrupt mode must be set to Mode 2.

Internal Structure

The internal structure of the Z80 PIO consists of a Z80 CPU bus interface, internal control logic, Port A I/O logic, Port B I/O logic, and interrupt control logic (Figure 4). The CPU bus interface logic allows the Z80 PIO to interface directly to the Z80 CPU with no other external logic. The internal control logic synchronizes the CPU data bus to the peripheral device interfaces (Port A and Port B). The two I/O ports (A and B) are virtually identical and are used to interface directly to peripheral devices.

Port Logic. Each port contains separate input and output registers, handshake control logic, and the control registers shown in Figure 5. All data transfers between the peripheral unit and the CPU use the data input and output registers. The handshake logic associated with each port controls the data transfers through the input and the output registers. The mode control register (two bits) selects one of the four programmable operating modes.

The control mode (Mode 3) uses the remaining registers. The input/output control register specifies which of the eight data bits in the port are to be outputs and enables these bits; the remaining bits are inputs. The mask register and the mask control register control Mode 3 interrupt conditions. The mask register specifies which of the bits in the port are active and which are masked or inactive.

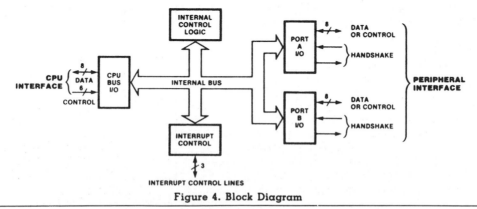

Figure 4. Block Diagram

Figure 3.9(c)

Internal Structure (Continued)

The mask control register specifies two conditions: first, whether the active state of the input bits is High or Low, and second, whether an interrupt is generated when any *one* unmasked input bit is active (OR condition) or if the interrupt is generated when *all* unmasked input bits are active (AND condition).

Interrupt Control Logic. The interrupt control logic section handles all CPU interrupt protocol for nested-priority interrupt structures. Any device's physical location in a daisy-chain configuration determines its priority. Two lines (IEI and IEO) are provided in each PIO to form this daisy chain. The device closest to the CPU has the highest priority. Within a PIO, Port A interrupts have higher priority than those of Port B. In the byte input, byte output, or bidirectional modes, an interrupt can be generated whenever the peripheral requests a new byte transfer. In the bit control mode, an interrupt can be generated when the peripheral status matches a programmed value. The PIO provides for complete control of nested interrupts. That is, lower priority devices may not interrupt higher priority devices that have not had their interrupt service routines completed by the CPU. Higher priority devices may interrupt the servicing of lower priority devices.

If the CPU (in interrupt Mode 2) accepts an interrupt, the interrupting device must provide an 8-bit interrupt vector for the CPU. This vector forms a pointer to a location in memory where the address of the interrupt service routine is located. The 8-bit vector from the interrupting device forms the least significant eight bits of the indirect pointer while the I Register in the CPU provides the most significant eight bits of the pointer. Each port (A and B) has an independent interrupt vector. The least significant bit of the vector is automatically set to 0 within the PIO because the pointer must point to two adjacent memory locations for a complete 16-bit address.

Unlike the other Z80 peripherals, the PIO does not enable interrupts immediately after programming. It waits until $\overline{M1}$ goes Low (e.g., during an opcode fetch). This condition is unimportant in the Z80 environment but might not be if another type of CPU is used.

The PIO decodes the RETI (Return From Interrupt) instruction directly from the CPU data bus so that each PIO in the system knows at all times whether it is being serviced by the CPU interrupt service routine. no other communication with the CPU is required.

CPU Bus I/O Logic. The CPU bus interface logic interfaces the Z80 PIO directly to the Z80 CPU, so no external logic is necessary. For large systems, however, address decoders and/or buffers may be necessary.

Internal Control Logic. This logic receives the control words for each port during programming and, in turn, controls the operating functions of the Z80 PIO. The control logic synchronizes the port operations, controls the port mode, port addressing, selects the read/write function, and issues appropriate commands to the ports and the interrupt logic. The Z80 PIO does not receive a write input from the CPU; instead, the $\overline{RD}$, $\overline{CE}$, C/D and $\overline{IORQ}$ signals generate the write input internally.

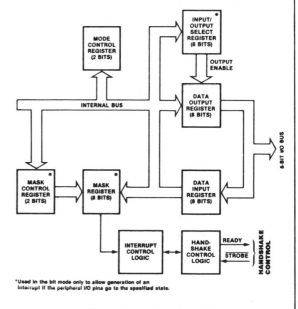

Figure 5. Typical Port I/O Block Diagram

Figure 3.9(d)

control register of the device by executing an output instruction in the software. Each port and the control register have separate addresses so it is a relatively simple task.

The Z80 PIO

The Z80 programmble I/O device contains a number of interesting features and it will therefore justify careful examination. The preceding pages (64–7) contain some extracts from its data sheet but because the device is so complex, only a few pages can be used. In all, the data sheet occupies 14 pages and includes all aspects of its operation such as its programming, the function of its pins, its timing diagrams and electrical a.c. and d.c. characteristics.

As can be seen from the first five pages of its data sheet, the Z80 PIO is a very complex device. Fortunately, it is relatively straightforward to use once the method of programming it has been established.

Main features

(a) The Z80 PIO operates in one of four modes which are programmed as the first byte to be sent to the port:
   (i)   Mode 0 – port acts as an output.
   (ii) Mode 1 – port acts as as input.
   (iii) Mode 2 – the port is bidirectional.
   (iv) Mode 3 – bit control mode.
   Each port A or B must be programmed into one of these modes to determine its basic operation.

(b) Each port has two handshake lines known as **strobe** and **ready**. These are connected if required between the PIO and the peripheral device to which the data is connected. The peripheral device must contain some logic which will acknowledge the **ready** signal from the PIO and return a **strobe** signal at the appropriate moment if synchronised data transfers are required. These are necessary in modes 0, 1 and 2 of operation, but they can be completely ignored in mode 3.

(c) Each port has a facility to interrupt the CPU,

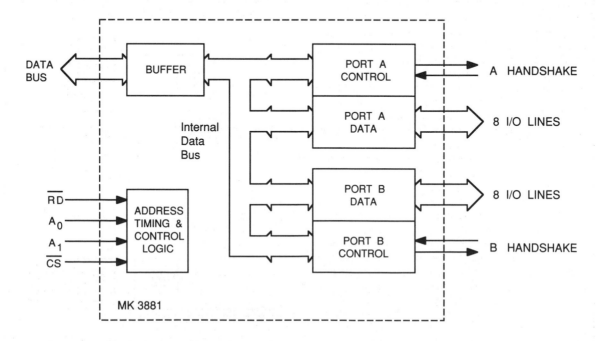

Figure 3.10   Simplified Z80 PIO internal structure

but this can be ignored if it is not required. This is controlled by interrupt control logic, which senses when specific data patterns pass through the port.

(d) All the control lines are connected to the Z80 microprocessor and are entirely compatible with its operation. No additional logic is required in a Z80 system.

(e) Because the PIO is programmable, it must be properly initialised before it can be used in a system. This is normally carried out as part of the start-up routine within the **monitor** program of the system.

Z80 PIO Programming

The Z80 PIO is a relatively complex chip which must be correctly programmed before it will function at all. *Figure 3.10* shows its internal structure. Although it contains only two 8-bit data ports, it also contains two control ports which control the operation of the data ports. The data ports are **initialised** by sending certain bytes to the control ports.

Generally, PIOs are arranged so that each of their four internal ports have separate consecutive addresses. For example the PIO may be allocated the addresses:

> Port A Data – 80 hex
> Port B Data – 81 hex
> Port A Control – 82 hex
> Port B Control – 83 hex

Thus to **initialise** Port A, the programmer must send the initialisation codes to Port 82 hex and, to initialise Port B, they must be sent to Port 83 hex. Each half of the PIO must be initialised separately, unlike some similar I/O devices such as the 8255, which can be completely initialised by sending only one byte to its control port.

Since the two halves of the PIO are almost identical, only Port A will be examined in further detail. The internal structure of Port A is shown in *Figure 3.11* (overleaf). The diagram shows both the control and data registers of Port A. The data port is designated by the double line surrounding the registers.

The function of each of the registers is as follows.

Data Port Registers

1. Output register This register latches output data from the CPU and sends it to any pin programmed to be an output.

2. Input register This register receives input data from external devices and may be read by the CPU at any time.

Control Registers

1. Mode control register The way in which the PIO functions is determined by the data programmed into this 2-bit register. There are four possible operating modes (*Table 3.1*).

Table 3.1 Possible operating modes for the PIO

Mode number	Mode register contents	Port operation
0	00	8 bits output
1	01	8 bits input
2	10	8 bits bidirectional
3	11	bit control (inputs/outputs)

Modes 0 and 1 are the simplest operating modes and both require the operation of the PIO handshake lines for successful operation. Mode 2 allows the PIO to operate both as an output and an input simultaneously so that it appears to be an extension of the system data bus. This is its bidirectional mode.

Mode 3 operation is by far the most versatile. In this mode each individual bit can be programmed to be either an input or an output. Thus the 8-bit port may comprise, for example, six input bits and two output bits.

2. Input/output select register Whenever mode 3 operation is programmed into the mode control register, the I/O select register must be programmed with a byte to determine which bits of the **data** port will be **inputs** and which bits will be **outputs**. A logic 1 indicates an **input** and a logic 0 indicate an **output**.

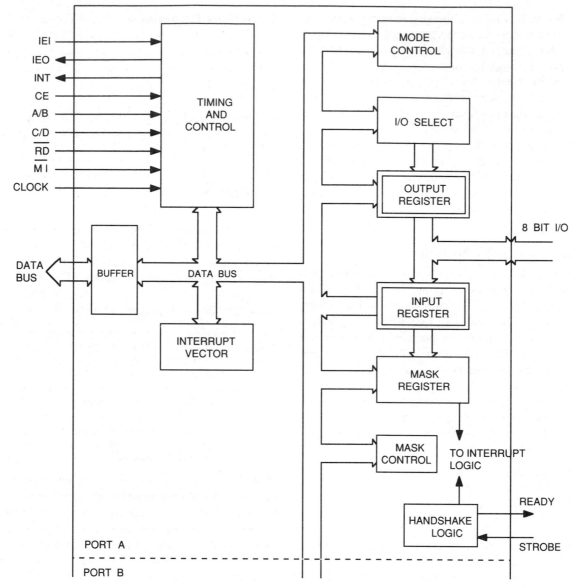

Figure 3.11 PIO port A structure

3. Mask control register This register is only required when the PIO will be used to generate an **interrupt**. It contains two bits that have specific functions. One determines whether an interrupt will be generated if the input signal is at a logic 0 or 1 and the other determines whether **AND** logic or **OR** logic will be applied to the input signals to generate the interrupt.

4. Mask register This register is programmed with a byte which determines which bits will be monitored for possible interrupting signals.

5. Interrupt vector Whenever **interrupts** are used, the Interrupt Vector register must be programmed with a number which will help to determine where an interrupt service subroutine will be found in memory.

The whole subject of interrupts will be dealt with in much more detail later in the book; so, for the moment, this can be ignored. Therefore, programming the PIO becomes a relatively simple matter of sending the appropriate commands to the mode control register and possibly the I/O select register.

PIO Programming

If the subject of interrupts is ignored, the following bytes of data must be sent to the control port for *each half* of the PIO.

1. Send the Mode Control Byte (*Figure 3.12*)

B_7 B_0

| M_1 | M_0 | X | X | 1 | 1 | 1 | 1 |

Figure 3.12 Z80 mode control bye

The two most significant bits set the **mode** of the data port. Bits 5 and 4 can be anything and bits 0–3 must all be logic 1.

For example, the program:

LD A,0FH
OUT (83H),A

would set up port B of the PIO into the **output** mode.

Similarly,

LD A,0FFH
OUT (82H),A

would set up Port A of the PIO into **bit control** mode, which is Mode 3.

2. When Mode 3 is selected, the I/O select byte must be sent next (*Figure 3.13*)

A logic 1 in any bit programs the corresponding bit of the data port to be an **input**. A logic 0 in any bit programs the corresponding bit of the data port to be an **output**.

| I/O$_7$ | I/O$_6$ | I/O$_5$ | I/O$_4$ | I/O$_3$ | I/O$_2$ | I/O$_1$ | I/O$_0$ |

Figure 3.13 Z80 I/O select byte

For example, to arrange bits 0–3 of port B as inputs and bits 4–7 to be outputs, send 0F hex:

LD A,0FH
OUT (83H),A

EXAMPLE

A certain system requires bits 0, 2, 3 and 7 of port A (80H) and bits 3, 4, 5 and 6 of port B to be **outputs**. All other bits should be inputs. Write a suitable initialisation program assuming the port numbers are as previously given (80H–83H).

The solution is as follows:

```
        ORG 1800H
INIT:   LD A,0FFH
        OUT (82H),A  ; SET PORT A TO MODE 3
        LD A,72H     ; 01110010 IN BINARY
        OUT (82H),A  ; SET UP A
        LD A,0FFH
        OUT (83H),A  ; SET PORT B TO MODE 3
        LD A,87H     ; 10000111 IN BINARY
        OUT (83H),A  ; SET UP B
```

3.6 PARALLEL-TO-SERIAL CONVERSION

Many computer applications require data to be transferred one bit at a time along a serial line. These include intercomputer data transfers over long distances, communication via telephone lines using modems and other connections between computers and peripherals such as printers. Since the computer data is in a parallel format, some means of converting this to a serial format is required.

The simplest method by which this can take place is to use a shift register which accepts parallel data but clocks it out in serial form. Many 74 series chips are capable of performing this function and these include the 74165, 74166, 74199. Each device performs parallel to serial conversion but their precise interface requirements and features vary slightly from one to the other. For example the 74199 is capable of shifting data either left or right in the register.

To connect the shift register as an output port

requires a small amount of interface logic which will allow the system to load the register using an output instruction. The parallel inputs of the shift register are connected to the data bus so that data can be written directly into the flip-flops that make up the register. In addition a clock circuit is required which will allow the data to be transmitted from the register at the required rate. Generally the clock circuit is made to be variable so that fixed data rates can be established according to the standard Baud rates. These are 300, 600, 1200, 2400, 4800, 9600, and 19200 bits per second respectively. Other higher Baud rates are also available for high-speed systems. *Figure 3.14* shows how a typical parallel load shift register could be connected as an output port in a system and *Figure 3.15* (pages 73–4) gives the precise data for one such device, the 74LS165.

Unfortunately there are a number of problems associated with the use of basic shift registers for serial-to-parallel conversion. These include:

(a) Additional logic circuits are required to control the operation of the shift register, and these must be specifically designed for each type of register used.

(b) Normal serial data connections in asynchronous mode use not only 8 bits but also a start and stop bit associated with each byte. This can be achieved with a shift register only by adding two extra flip-flops one at each end of the register. This clearly involves additional circuit design.

(c) The system is relatively inflexible since it is not possible with the hardware design to change from 8 to 7 data bits with ease.

(d) A shift register has no built-in mechanism to inform the processor when the data has all been transmitted. This means that some timing must take place within the software to ensure that data is not written into the register before the previous byte has been transmitted. This leads to software overheads and delays.

(e) If the shift register must transmit data to a modem, there are no facilities within the device to control the operation of the modem and these must be built with additional circuits.

Despite all of these limitations shift registers are still frequently used to perform parallel-to-serial conversion and transmission in microcomputer systems, but they tend to be used in fixed applications which are not likely to have to be changed.

A far more common method of performing parallel-to-serial conversion is to use a dedicated chip known as an **asynchronous receiver/transmitter** or a **serial input/output device**.

There are two such devices in the Z80 family of

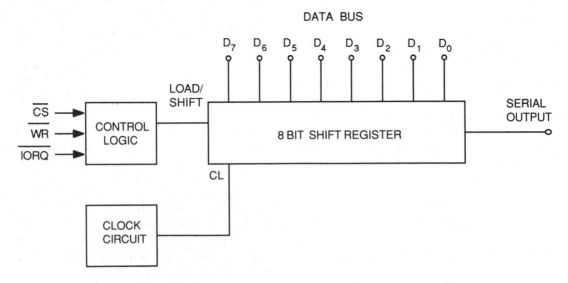

Figure 3.14 Serial to parallel output with a shift register

TTL
MSI

TYPES SN54165, SN54LS165A, SN74165, SN74LS165A
PARALLEL-LOAD 8-BIT SHIFT REGISTERS

- Complementary Outputs
- Direct Overriding Load (Data) Inputs
- Gated Clock Inputs
- Parallel-to-Serial Data Conversion

TYPE	TYPICAL MAXIMUM CLOCK FREQUENCY	TYPICAL POWER DISSIPATION
'165	26 MHz	210 mW
'LS165A	35 MHz	105 mW

SN54165, SN54LS165A . . . J OR W PACKAGE
SN74165, SN74LS165A . . . J OR N PACKAGE
(TOP VIEW)

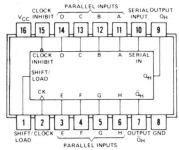

positive logic: see description

description

The '165 and 'LS165A are 8-bit serial shift registers that shift the data in the direction of Q_A toward Q_H when clocked. Parallel-in access to each stage is made available by eight individual direct data inputs that are enabled by a low level at the shift/load input. These registers also feature gated clock inputs and complementary outputs from the eighth bit. All inputs are diode-clamped to minimize transmission-line effects, thereby simplifying system design.

Clocking is accomplished through a 2-input positive-NOR gate, permitting one input to be used as a clock-inhibit function. Holding either of the clock inputs high inhibits clocking and holding either clock input low with the shift/load input high enables the other clock input. The clock-inhibit input should be changed to the high level only while the clock input is high. Parallel loading is inhibited as long as the shift/load input is high. Data at the parallel inputs are loaded directly into the register on a high-to-low transition of the shift/load input independently of the levels of the clock, clock inhibit, or serial inputs.

logic symbol

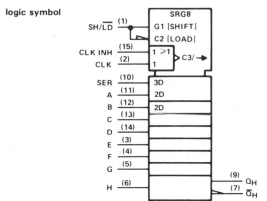

FUNCTION TABLE.

INPUTS					INTERNAL OUTPUTS		OUTPUT
SHIFT/ LOAD	CLOCK INHIBIT	CLOCK	SERIAL	PARALLEL A . . . H	Q_A	Q_B	Q_H
L	X	X	X	a . . . h	a	b	h
H	L	L	X	X	Q_{A0}	Q_{B0}	Q_{H0}
H	L	↑	H	X	H	Q_{An}	Q_{Gn}
H	L	↑	L	X	L	Q_{An}	Q_{Gn}
H	H	X	X	X	Q_{A0}	Q_{B0}	Q_{H0}

See explanation of function tables on page 3-8.

schematic of inputs and output

'165

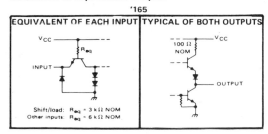

'LS165A

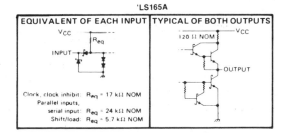

Figure 3.15(a)–(b)   74LS165 data sheet (Courtesy Texas Instruments)

TYPES SN54165, SN54LS165A, SN74165, SN74LS165A
PARALLEL-LOAD 8-BIT SHIFT REGISTERS

functional block diagram

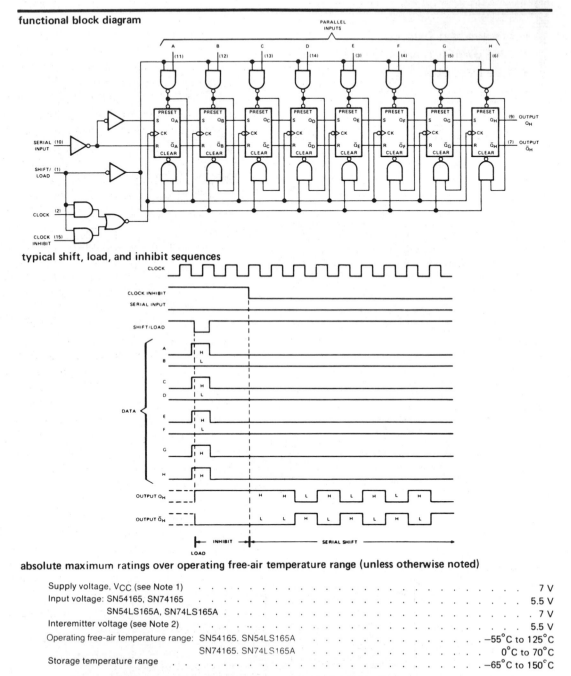

typical shift, load, and inhibit sequences

absolute maximum ratings over operating free-air temperature range (unless otherwise noted)

Supply voltage, V_{CC} (see Note 1) . . . . 7 V
Input voltage: SN54165, SN74165 . . . . 5.5 V
            SN54LS165A, SN74LS165A . . . . 7 V
Interemitter voltage (see Note 2) . . . . 5.5 V
Operating free-air temperature range: SN54165. SN54LS165A . . . . −55°C to 125°C
                      SN74165. SN74LS165A . . . . 0°C to 70°C
Storage temperature range . . . . −65°C to 150°C

NOTES: 1. Voltage values, except interemitter voltage, are with respect to network ground terminal.
        2. This is the voltage between two emitters of a multiple-emitter transistor. This rating applies for the '165 to the shift/load input in conjunction with the clock-inhibit inputs.

Figure 3.15(b)

chips. One is a **dual asynchronous receiver/transmitter** which provides two asynchronous receiver/transmitter ports within one chip and includes control for the modems which may or may not be connected. The other is a **serial input/output controller** and again provides two independent asynchronous or synchronous channels with complete modem control. Synchronous data transmission is generally of a higher speed than asynchronous transmission and the SIO chip includes a number of different serial synchronous protocols, together with the necessary automatic error checking.

Synchronous systems generally omit any synchronisation pulses between bytes of data in each block and this allows data to be transmitted faster.

The simplest device to consider is the asynchronous receiver/transmitter.

3.7 Z80 DUAL ASYNCHRONOUS RECEIVER/TRANSMITTER

The complete data sheets for the Z80 DART are too complex to reproduce in full here, but the first three pages give a general description of the device together with its pin connections and pin descriptions (*Figure 3.16*, pages 76–8).

The Z80 DART provides two independent asynchronous channels within a single device, known as **Channel A** and **Channel B**. Each channel represents two ports in the Z80 system and these are selected with the pins B/A and C/D which would normally be connected to address lines A_0 and A_1 respectively. This would give the following port allocations:

  Port 00 – Channel A data
  Port 01 – Channel B data
  Port 02 – A Control
  Port 03 – B Control

Whenever the device is used, information must be pre-programmed into internal registers to initialise it, and this must be done for both transmit and receive operations. The control registers are shown in *Figure 3.17* (page 79) with the read/write registers.

Because the device is designed to complement the Z80 microprocessor, its interface with the CPU utilises the standard Z80 control signals direct from the microprocessor. These are also shown on the first page of the data sheet as the CPU control signals.

In addition it also has the standard Z80 interrupt signal so that this device can be used with other Z80 peripheral chips in a relatively complex interrupt structure if required.

Figure 3.17 also shows the interface between the DART and two serial channels, Channel A and Channel B. Each channel is an independent transmit and receive channel with independent clock signals and other modem control signals. Each channel is a full duplex channel (which means that it can transmit and receive simultaneously) and it is intended that it drives an RS232 line. The modem control signals include the usual request to send, clear to send, and data carrier detect lines, together with a ring indicator output.

The device can be used to transmit or receive between 5 and 8 bits per character with optional party bits, either even or odd, and 1, 1.5 or 2 stop bits. Automatic error detection is also included for framing errors and overrun errors.

The internal architecture of each channel of the DART is shown in *Figure 3.18* (page 79). Incoming data appears at the RxDA input, and it is clocked into the 8-bit receive shift register. It is then loaded out in parallel into three first-in first-out registers until it is available to be read on the data bus. This three register delay provides an additional time for the microprocessor to respond to an interrupt when high-speed data transfers are in operation.

Data to be transmitted is written into the transmit data register and it is then clocked into a 9-bit shift register. This effectively adds start and stop bits to the data which is then transmitted via the TxDA output. Data transmission and reception are controlled by the speed of the transit clock and receive clock respectively, and are independent of the microprocessor clock speed.

DART programming

Each channel of the Z80 DART has nine control registers, some of which are read registers and some which are write registers. These must be

Dual Asynchronous Receiver/Transmitter

- Two independent full-duplex channels with separate modem controls. Modem status can be monitored.

- Receiver data registers are quadruply buffered; the transmitter is doubly buffered.

- Interrupt features include a programmable interrupt vector, a "status affects vector" mode for fast interrupt processing, and the standard Z80 peripheral daisy-chain interrupt structure that provides automatic interrupt vectoring with no external logic.

- In x1 clock mode, data rates are 0 to 500'< bits/second with a 2.5 MHz clock or 0 to 800K bits/second with a 4.0 MHz clock, or 0 to 1200K bit/second with a 6.0 MHz clock.

- Programmable options include 1, 1½ or 2 stop bits; even, odd or no parity; and x1, x16, x32 and x64 clock modes.

- Break generation and detection as well as parity-, overrun- and framing-error detection are available.

General Description

The Z80 DART (Dual-Channel Asynchronous Receiver/Transmitter) is a dual-channel multi-function peripheral component that satisfies a wide variety of asynchronous seral data communications requirements in microcomputer systems. The Z80 DART is used as a serial-to-parallel, parallel-to-serial converter/controller in asynchronous applications. In addition, the device also provides modem controls for both channels. In applications where modem controls are not needed, these lines can be used for general-purpose I/O.

The Z80 SIO, a more versatile device, provides synchronous (Bisync, HDLC and SDLC) as well as asynchronous operation.

The Z80 DART is fabricated with n-channel silicon-gate depletion-load technology, and is packaged in a 40 pin plastic or ceramic DIP.

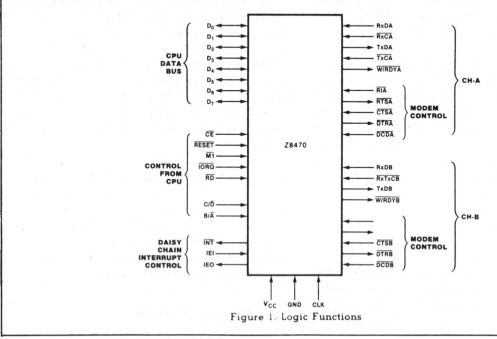

Figure 1. Logic Functions

Figure 3.16(a)–(c) Z80 DART data sheet (Courtesy SGS)

General Description (Continued)

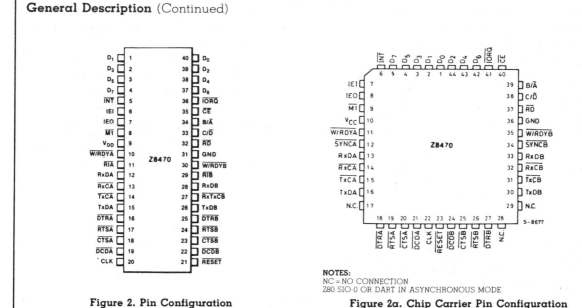

Figure 2. Pin Configuration

Figure 2a. Chip Carrier Pin Configuration

NOTES:
NC = NO CONNECTION
Z80 SIO-0 OR DART IN ASYNCHRONOUS MODE

Pin Description

B/Ā *Channel A Or B Select* (input, High selects Channel B). This input defines which channel is accessed during a data transfer between the CPU and the Z80 DART

C/D̄ *Control Or Data Select* (input, High selects Control). This input specifies the type of information (control or data) transferred on the data bus between the CPU and the Z80 DART.

CE *Chip Enable* (input, active Low). A Low at this input enables the Z80 DART to accept command or data input from the CPU during a write cycle, or to transmit data to the CPU during a read cycle.

CLK. *System Clock* (input). The Z80 DART uses the standard Z80 single-phase system clock to synchronize internal signals.

CTSA, CTSB. *Clear To Send* (inputs, active Low). When programmed as Auto Enables, a Low on these inputs enables the respective transmitter. If not programmed as Auto Enables, these inputs may be programmed as general-purpose inputs. Both inputs are

Schmitt-trigger buffered to accomodate slow-risetime signals.

D₀-D₇. *System Data Bus* (bidirectional, 3-state) transfers data and commands between the CPU and the Z80 DART.

DCDA, DCDB. *Data Carrier Detect* (inputs, active Low). These pins function as receiver enables if the Z80 DART is programmed for Auto Enables; otherwise they may be used as general-purpose input pins. Both pins are Schmitt-trigger buffered.

DTRA, DTRB. *Data Terminal Ready* (outputs, active Low). These outputs follow the state programmed into the DTR bit. They can also be programmed as general-purpose outputs.

IEI. *Interrupt Enable In* (input, active High) is used with IEO to form a priority daisy chain when there is more than one interrupt-driven device. A High on this line indicates that no other device of higher priority is being serviced by a CPU interrupt service routine.

Figure 3.16(b)

Pin Description (Continued)

IEO. *Interrupt Enable Out* (output, active High). IEO is High only if IEI is High and the CPU is not servicing an interrupt from this Z80 DART. Thus, this signal blocks lower priority devices from interrupting while a higher priority device is being serviced by its CPU interrupt service routine.

INT. *Interrupt Request* (output, open drain, active Low). When the Z80 DART is requesting an interrupt, it pulls $\overline{\text{INT}}$ Low.

M1. *Machine Cycle One* (input from Z80 CPU, active Low). When $\overline{\text{M1}}$ and $\overline{\text{RD}}$ are both active, the Z80 CPU is fetching an instruction from memory; when $\overline{\text{M1}}$ is active while $\overline{\text{IORQ}}$ is active, the Z80 DART accepts $\overline{\text{M1}}$ and $\overline{\text{IORQ}}$ as an interrupt acknowledge if the Z80 DART is the highest priority device that has interrupted the Z80 CPU.

IORQ. *Input/Output Request* (input from CPU, active Low). $\overline{\text{IORQ}}$ is used in conjunction with B/$\overline{\text{A}}$, C/$\overline{\text{D}}$, $\overline{\text{CE}}$ and $\overline{\text{RD}}$ to transfer commands and data between the CPU and the Z80 DART. When $\overline{\text{CE}}$, $\overline{\text{RD}}$ and $\overline{\text{IORQ}}$ are all active, the channel selected by B/$\overline{\text{A}}$ transers data to the CPU (a read operation). When $\overline{\text{CE}}$ and $\overline{\text{IORQ}}$ are active, but $\overline{\text{RD}}$ is inactive, the channel selected by B/$\overline{\text{A}}$ is written to by the CPU with either data or control information as specified by C/$\overline{\text{D}}$.

$\overline{\text{R}\times\text{CA}}$, $\overline{\text{R}\times\text{CB}}$. *receiver Clocks* (inputs). Receive data is sampled on the rising edge of $\overline{\text{R}\times\text{C}}$. The Receive Clocks may be 1, 16, 32 or 64 times the data rate.

$\overline{\text{RD}}$. *Read Cycle Status* (input from CPU, active Low). If $\overline{\text{RD}}$ is active, a memory or I/O read operation is in progress.

R×DA, R×DB. *Receive Data* (inputs, active High).

RESET. *Reset* (input, active Low). Disables both receivers and transmitters, forces T×DA and T×DB marking, forces the modem controls High and disables all interrupts.

$\overline{\text{RIA}}$, $\overline{\text{RIB}}$. *Ring Indicator* (inputs, Active Low). These inputs are similar to $\overline{\text{CTS}}$ and DCD. The Z80 DART detects both logic level transitions and interrupts the CPU. When not used in switched-line applications, these inputs can be used as general-purpose inputs.

$\overline{\text{RTSA}}$, $\overline{\text{RTSB}}$. *Request to Send* (outputs, active Low). When the RTS bit is set, the $\overline{\text{RTS}}$ output goes Low. When the RTS bit is reset, the output goes High after the transmitter empties.

$\overline{\text{T}\times\text{CA}}$, $\overline{\text{T}\times\text{CB}}$. *Transmitter Clocks* (inputs). T×D changes on the falling edge of $\overline{\text{T}\times\text{C}}$. The Transmitter Clocks may be 1, 16, 32 or 64 times the data rate; however, the clock multiplier for the transmitter and the receiver must be the same. The Transmitt Clock inputs are Schmitt-triggered buffered. Both the Receiver and Transmitter Clocks may be driven by the Z80 CTC Counter Time Circuit for programmable baud rate generation.

T×DA, T×DB. *Transmitt Data* (outputs, active High).

$\overline{\text{W}}$/RDYA, $\overline{\text{W}}$/RDYB. *Wait/Ready* (outputs, open drain when programmed for Wait function, driven High and Low when programmed for Ready function). These dual-purpose outputs may be programmed as Ready lines for a DMA controller or as Wait lines that sychronize the CPU to the Z80 DART data rate. The reset state is open drain.

Figure 3.16(c)

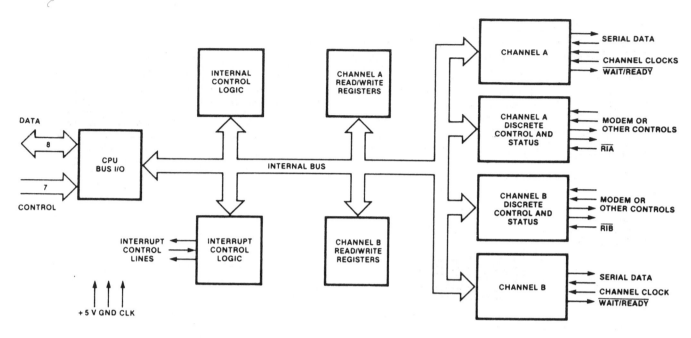

Figure 3.17 Z80 DART block diagram (Courtesy SGS)

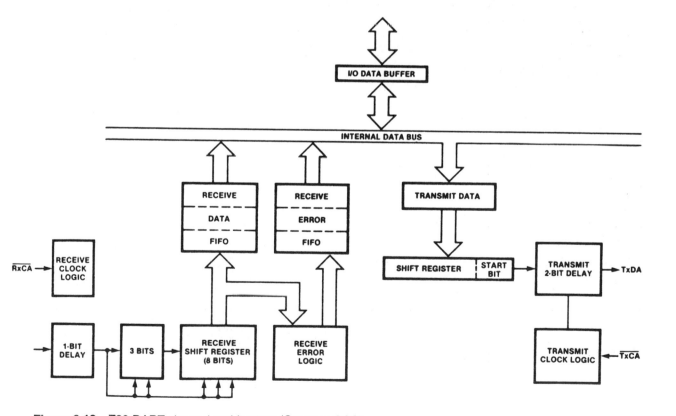

Figure 3.18 Z80 DART channel architecture (Courtesy SGS)

Z80 DART Read and Write Registers

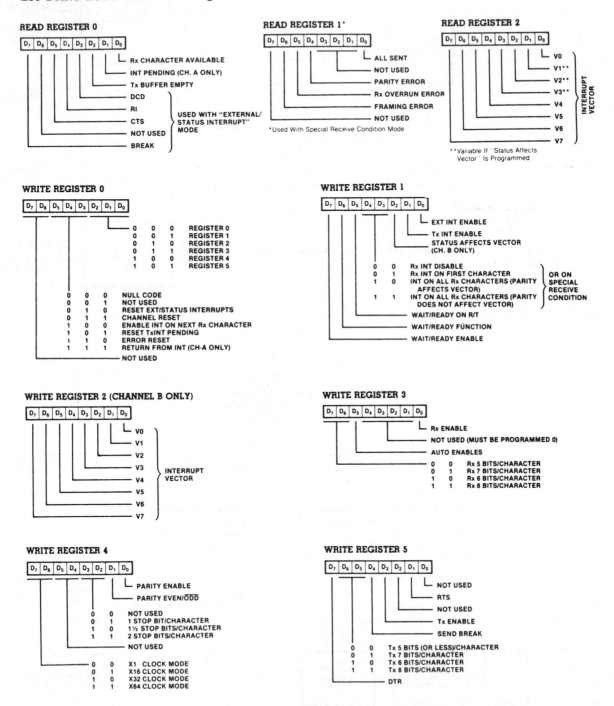

Figure 3.19 Z80 DART read and write registers (Courtesy SGS)

programmed before the device can be operated, so that it is correctly configured for the type of operation required. A diagram of the registers is shown in *Figure 3.19*.

The complexity of each of the registers can be seen from the figure and it is therefore not intended to fully explore the method of programming this particular device. Only register 0 can be read or written to with a single byte, written to the control port of the respective channel. Other registers must first have data written into write register 0 with the appropriate register numbers specified in bits 0, 1 and 2, then data is written into the appropriate register.

The main functions of the registers are as follows.

Read Registers

RR0   Transmit/receive buffer status, interrupt status and external status.

RR1   Special receive condition status.

RR2   Modified interrupt vector channel B only.

Write Register Functions

WR0   Register pointers, initialisation commands for the various modes, etc.

WR1   Transmit/receive interrupt and data transfer mode definition.

WR2   Interrupt vector (Channel B only).

WR3   Receive parameters and controls.

WR4   Transmit/receive miscellaneous parameters and modes.

WR5   Transit parameters and controls.

Because of the relatively large number of registers which need to be programmed in the Z80 DART, it can be seen that a relatively long initialisation program would be required for setting up both channels to transmit and receive data. However, for simple operation only a few of the registers need data to be written to them.

The subject of serial data transmission can be relatively simple, with the use of a basic shift register to perform the parallel-to-serial conversion, or it can be very complex with the use of a dedicated serial I/O chip. Each method can be made to produce the same result, which is a serial stream of data preceded by a start bit, generally a logic 0 and followed by one or more stop bits at the logic 1 level. When this is sent as an RS232 signal, it is translated to $+12$ and -12 V to represent a logic 0 and a logic 1 level respectively. Thus the data is effectively transmitted 'upside down' but the original levels are of course restored when the data is received.

Summary

This chapter has considered a number of VLSI and other microprocessor system devices. These form the heart of modern systems and help to reduce their chip count and increase their reliability:

- Microprocessors are the most complex devices likely to be found in most systems and this is reflected in the size of their data sheets. Typically a microprocessor can only be specified completely in a small book.

- Microprocessor data sheets must specify every aspect of the hardware and software characteristics of the device.

- Parallel I/O can be provided with simple MSI latches and buffers, but these are of restricted use.

- Programmable I/O devices are very flexible in operation since they operate according to their software initialisation routines and not hardware considerations. They require comprehensive manufacturer's data.

- Parallel-to-serial conversion can be carried out by MSI shift registers, but full serial data transmission systems would normally require many more logic circuits for a complete implementation.

- Comprehensive serial input/output facilities can be provided by a dedicated SIO or DART device.

Questions

3.1 Is the 68000 microprocessor fully TTL compatible?

3.2 Why does the 68000 appear to have no I/O request line?

3.3 Compare the two addressing ranges of the Z80 and 68000, and relate this to the number of address lines.

3.4 Calculate the maximum power supply current, if the 68000 is dissipating its maximum power.

3.5 List the main advantages and disadvantages of using a dedicated serial I/O chip rather than a standard shift register to perform serial data transfers.

3.6 In a Z80 DART, explain the difference between the inputs CLK (system clock) and TxCA (transmitter clock).

3.7 A Z80 PIO with port numbers 80H–83H must be programmed so that Port A is an input port and Port B has bits 0–3 as inputs and bits 4–7 as outputs. Write a suitable initialisation program assuming that no interrupt facilities are required, and that Mode 3 operation is required for both ports.

3.8 What is the function of the mask control register in a Z80 PIO?

FOUR

Memory devices

When you have completed this chapter, you should be able to:

1. Discuss, using manufacturer's literature, the function, operation and distinguishing characteristics of:

   (a) Static RAM.
   (b) Dynamic RAM.
   (c) MOS ROM.
   (d) EPROM.
   (e) EEPROM.

4.1 STATIC RAM

There are two main types of memory in most microcomputer systems, these are **RAM** and **ROM**. Random Access Memory (RAM) comes in two types known as **static** and **dynamic**.

RAM devices can have data written into them or read from them, and retain data only while the power is applied. They are therefore used to store information which needs to be changed frequently. They generally have very rapid read/write times and this may typically be between 40 nS and 250 nS.

Static RAM devices are based on a two transistor memory cell and therefore need no refresh facilities like dynamic devices. However this has a number of disadvantages and in general, static RAM chips have a higher power dissipation than dynamic devices. A typical static RAM cell is shown in *Figure 4.1* (overleaf). The basic RAM consists of two transistors with cross-coupled gates and the other two transistors are used to read data in and out of the memory.

The range of static RAMs available to the designer is considerable. There are some small high speed memories with less than 1K capacity and others with $2K \times 8$ capacity and relatively lower speed. Most static RAMs require only a single $+5$ V supply and this fact alone makes them easy to integrate within microcomputer systems. Typically the organisation of static RAMs is to have either 4 or 8 bits per chip. The number of bits per chip determines the package size which may be either 16, 18, 20, 22 or 24 pins.

MK 4118 Static RAM

One typical device from the range is the MOSTEK MK4118 device. This is a 1024×8 bit device, which comes in a 24 pin dual-in-line package.

Its data sheets are very comprehensive and are given on the next few pages in their entirety which should help to provide some idea of the detail to which manufacturers must go to specify complex circuits.

This data sheet follows the pattern of most VLSI data sheets with a summary of the main features of the device first followed by a description of the main points, with a block diagram and pin connection diagram. This is then followed by the main electrical operating conditions including its recommended d.c. operating conditions and its typical characteristics. The main a.c. electrical characteristics are related to its timing functions and in particular relate to the diagrams in *Figure 4.2* (pages 85–9) where each rise time, fall time and delay is specified very precisely. Memory devices may operate in a number of modes and the operation of each mode is described in detail. This allows the designer to design circuits using the memories whose timing can be made to match

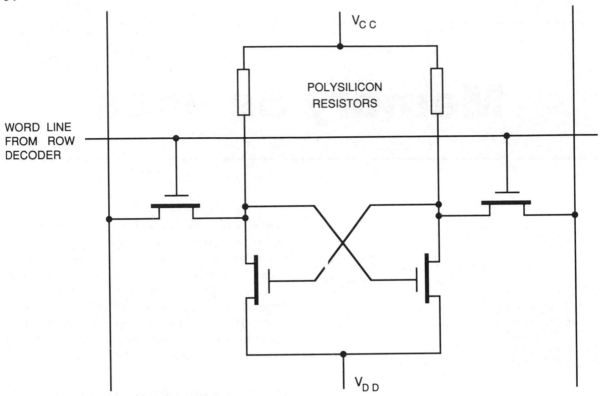

Figure 4.1 Static memory cell

those of other devices in the system. Synchronisation is very important. The examination of every aspect of this and other circuits to be considered is much too complex. However, some of the main features are worth noting.

Main features

(a) The 4118 operates from a single +5 V supply and comes in a 24 pin package which is compatible with other ROM/PROM devices. Its power dissipation at 400 mW is relatively low.

(b) The part number may be used to specify the performance in terms of the access time of the device, such as MK4118–1, MK4118–2. Here the higher the number, the higher the access time. Memory devices are not designed for specific access times in general, but they are tested after manufacture and then selected and stamped with the appropriate number so that the fastest devices can be sold at a higher price than the slower ones.

(c) The input and output pins are totally TTL compatible with a typical fan-out of 12 LSTTL devices. This means that the outputs of the device are perfectly capable of driving a data bus which may have TTL devices and others connected to it.

(d) The names of the pins given in the pin connection diagram are not the same as those you might expect from a microprocessor pin connection diagram. There are some similarities, but these are limited to the address lines and the chip select $\overline{CS}$. Even the ground line is labelled V_{SS}.

Some help is afforded by the pin names given below the pin connection diagram. The data I/O pins which would be described as the data bus in a microprocessor are labelled I/O1–I/O8. Note that these are numbers 1–8 rather than D_0–D_7 as they would be in a microprocessor diagram.

1K x 8-BIT STATIC RAM
MK4118(P/N)Series

FEATURES

☐ Address Activated ™ Interface combines benefits of Edge Activated ™ and fully static.

☐ High performance

Part number	Access time	Cycle time
MK4118-1	120 nsec	120 nsec
MK4118-2	150 nsec	150 nsec
MK4118-3	200 nsec	200 nsec
MK4118-4	250 nsec	250 nsec

☐ Single +5 volt power supply

☐ TTL compatible I/O

    Fanout: 2 - Standard TTL
               2 - Schottky TTL
            12 - Low power Schottky TTL

☐ Low Power - 400mw Active

☐ 24-pin ROM/PROM compatible pin configuration

☐ $\overline{CS}$, $\overline{OE}$, and $\overline{LATCH}$ functions for flexible system operation

☐ Read-Modify-Write Capability

DESCRIPTION

The MK4118 uses MOSTEK's Poly R N-Channel Silicon Gate process and advanced circuit design techniques to package 8192 bits of static RAM on a single chip. MOSTEK's address activated ™ circuit design technique is utilized to achieve high performance, low power, and easy user implementation. The device has a V_{IH} = 2.2, V_{IL} = 0.8V, V_{OH} = 2.4, V_{OL} = 0.4V making it totally compatible with all TTL family devices.

The MK4118 is designed for all wide word memory applications. The MK4118 provides the user with a high-density, cost-effective 1Kx8 bit Random Access Memory. Fast Output Enable ($\overline{OE}$) and Chip Select ($\overline{CS}$) controls are provided for easy interface in microprocessor or other bus-oriented systems. The MK4118 features a flexible Latch ($\overline{L}$) function to permit latching of the address and $\overline{CS}$ status at the user's option. Common data and address bus operation may be performed at the system level by utilizing the $\overline{L}$ and $\overline{OE}$ functions for the MK4118. The latch function may be bypassed by merely tying the latch pin to V_{CC}, providing fast ripple-through operation.

BLOCK DIAGRAM

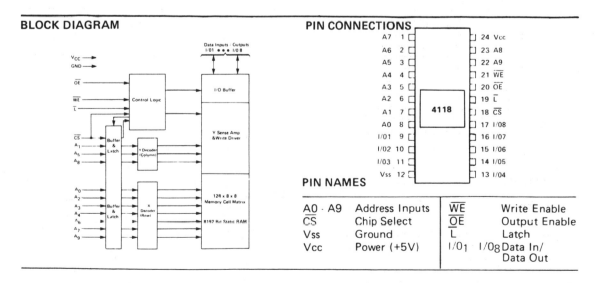

PIN NAMES

A0 - A9	Address Inputs	$\overline{WE}$	Write Enable
$\overline{CS}$	Chip Select	$\overline{OE}$	Output Enable
Vss	Ground	L	Latch
Vcc	Power (+5V)	I/O_1 - I/O_8	Data In/ Data Out

Figure 4.2(a)–(e) MK4118 data sheets (Courtesy Mostek)

ABSOLUTE MAXIMUM RATINGS*

Voltage on any pin relative to V_{SS} . . . . −0.5V to +7.0V

Operating Temperature T_A (Ambient) . . . . 0° C to + 70° C

Storage Temperature (Ambient) (Ceramic) . . . . −65° C to +150° C

Storage Temperature (Ambient) (Plastic) . . . . −55° C to +125° C

Power Dissipation . . . . 1 Watt

Short Circuit Output Current 20mA

*Stresses greater than those listed under "Absolute Maximum Ratings" may cause permanent damage to the device. This is a stress rating only and functional operation of the device at these or any other conditions above those indicated in the operational sections of this specification is not implied. Exposure to absolute maximum rating conditions for extended periods may affect reliability

RECOMMENDED DC OPERATING CONDITIONS [3]
(0° C ⩽ T_A ⩽ + 70° C)

	PARAMETER	MIN	TYP	MAX	UNITS	NOTES
V_{CC}	Supply Voltage	4.75	5 0	5.25	Volts	1
V_{SS}	Supply Voltage	0	0	0	Volts	1
V_{IH}	Logic "1" Voltage All Inputs	2.2		7.0	Volts	1
V_{IL}	Logic "0" Voltage All Inputs	−0.3		0.8	Volts	1

DC ELECTRICAL CHARACTERISTICS [1,3]
(0°C ⩽ T_A ⩽ + 70°C) (V_{CC} = 5.0 volts ⁼ 5%)

	PARAMETER	MIN	MAX	UNITS	NOTES
I_{CC1}	Average V_{CC} Power Supply Current (Active)		80	mA	
I_{CC2}	Average V_{CC} Power Supply Current (Standby)		60	mA	5
I_{IL}	Input Leakage Current (Any Input)	−10	10	μA	2
I_{OL}	Output Leakage Current	−10	10	μA	2
V_{OH}	Output Logic "1" Voltage I_{OUT}= −1mA	2.4		Volts	
V_{OL}	Output Logic "0" Voltage I_{OUT}= 4mA		0.4	Volts	

AC ELECTRICAL CHARACTERISTICS [1,3]
(0° C ⩽ T_A ⩽ + 70°C) (V_{CC} = + 5.0 volts ± 5%)

	PARAMETER	TYP	MAX	NOTES
C_I	Capacitance on all pins except I/O	4pF		4
$C_{I/O}$	Capacitance on I/O pins	10pF		4

NOTES:
1. All voltages referenced to V_{SS}.
2. Measured with $0 ⩽ V_I ⩽ 5V$ and outputs deselected (V_{cc} = 5V)
3. A minimum of 100 μsec time delay is required after application of VCC (+5V) before proper device operation can be achieved.
4. Effective capacitance calculated from the equation $C = I \frac{\Delta t}{\Delta V}$ with ΔV = 3V and V_{CC} nominal
5. Standby mode is defined as condition when addresses, latch and $\overline{WE}$ remain unchanged.

OPERATION

READ MODE

The MK4118 is in the READ MODE whenever the Write Enable control input ($\overline{WE}$) is in the high state. The state of the 8 data I/O signals is controlled by the Chip Select ($\overline{CS}$) and Output Enable ($\overline{OE}$) control sig-nals. The READ MODE memory cycle may be either STATIC (ripple-through) or LATCHED, depending on user control of the Latch Input Signal ($\overline{L}$).

STATIC READ CYCLE

In the STATIC READ CYCLE mode of operation, the MK4118 provides a fast address ripple-through

Figure 4.2(b)

ELECTRICAL CHARACTERISTICS [6]
(0°C ≤ T$_A$ ≤ 70°C and V$_{CC}$ = 5.0 volts ± 5%)

SYMBOL	PARAMETER	MK4118-1		MK4118-2		MK4118-3		MK4118-4		UNIT	NOTE
		MIN	MAX	MIN	MAX	MIN	MAX	MIN	MAX		
t$_{RC}$	Read Cycle Time	120		150		200		250		ns	
t$_{AA}$	Address Access Time		120		150		200		250	ns	
t$_{CSA}$	Chip Select Access Time		60		75		100		125	ns	
t$_{CSZ}$	Chip Select Data Off Time	0	60	0	75	0	100	0	125	ns	
t$_{OEA}$	Output Enable Access Time		60		75		100		125	ns	
t$_{OEZ}$	Output Enable Data Off Time	0	60	0	75	0	100	0	125	ns	
t$_{AZ}$	Address Data Off Time	10		10		10		10		ns	
t$_{ASL}$	Address To Latch Setup Time	10		10		10		20		ns	
t$_{AHL}$	Address From Latch Hold Time	40		50		65		80		ns	
t$_{CSL}$	$\overline{CS}$ To Latch Setup Time	0		0		0		0		ns	
t$_{CHL}$	$\overline{CS}$ From Latch Hold Time	40		50		65		80		ns	
t$_{LA}$	Latch Off Access Time		155		200		260		320	ns	
t$_{WC}$	Write Cycle Time	120		150		200		250		ns	
t$_{ASW}$	Address To Write Setup Time	0		0		0		0		ns	
t$_{AHW}$	Address From Write Hold Time	40		50		65		80		ns	
t$_{CSW}$	$\overline{CS}$ To Write Setup Time	0		0		0		0		ns	
t$_{CHW}$	$\overline{CS}$ From Write Hold Time	40		50		65		80		ns	
t$_{DSW}$	Data To Write Setup Time	20		30		40		50		ns	
t$_{DHW}$	Data From Write Hold Time	20		30		40		50		ns	
t$_{WD}$	Write Pulse Duration	35		50		60		70		ns	
t$_{LDH}$	Latch Duration, High	35	DC	50	DC	60	DC	70	DC	ns	
t$_{LDL}$	Latch Duration, Low		DC		DC		DC		DC	ns	
t$_{WEZ}$	Write Enable Data Off Time	0	60	0	75	0	100	0	125	ns	
t$_{LZ}$	Latch Data Off Time	10		10		10		10		ns	
t$_{WPL}$	Write Pulse Lead Time	75		90		130		170			

NOTES:
6. AC timing measurements made with 2 TTL loads plus 100pF.

STATIC READ CYCLE (Cont'd)

access of data from 8 of 8192 locations in the static storage array. Thus, the unique address specified by the 10 Address Inputs (An) define which 1 of 1024 bytes of data is to be accessed. The STATIC READ CYCLE is defined by $\overline{WE}$ = L = High.

A transition on any of the 10 address inputs will disable the 8 Data Output Drivers after t$_{AZ}$. Valid Data will be available to the 8 Data Output Drivers within t$_{AA}$ after all address input signals are stable, and the data will be output under control of the Chip Select ($\overline{CS}$) and Output Enable ($\overline{OE}$) signals.

LATCHED READ CYCLE

The LATCHED READ CYCLE is also defined by the Write Enable control input ($\overline{WE}$) being in the high state, and it is synchronized by proper control of the Latch ($\overline{L}$) input.

As the Latch control input ($\overline{L}$) is taken low, Address (An) and Chip Select ($\overline{CS}$) inputs that are stable for the specified set-up and hold times are latched internally. Data out corresponding to the latched address will be supplied to the Data Output drivers. The output drivers will be enabled to drive the Output Data Bus under control of the Output Enable ($\overline{OE}$) and latched Chip Select ($\overline{CS}$) inputs.

Taking the latch input high begins another read cycle for the memory locations specified by the address then appearing on the Address Input (An). Returning the latch control to the low state latches the new Address and Chip Select inputs internally for the remainder of the LATCHED READ CYCLE.

Figure 4.2(c)

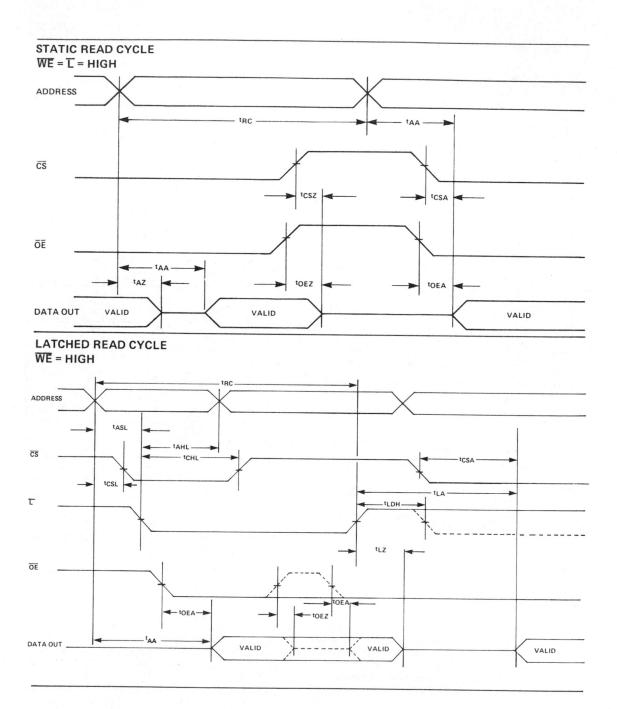

Figure 4.2(d)

WRITE CYCLE
OE = LOW, L = HIGH

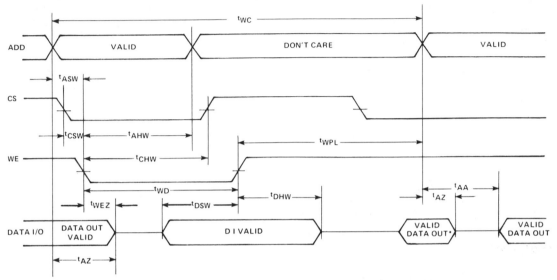

*NOTE: Assumes $t_{WC} \geqslant t_T + t_{WD}(min) + t_{WPL}(min)$. If not then output will remain open as $t_{AZ} < t_{AA}$ for an address change. Also $\overline{OE}$ may be used to maintain DO open.

LATCHED READ CYCLE (Cont'd)

NOTE: If 'L' pin is left open it will automatically assume the 'high' state.

WRITE MODE

The MK4118 is in the WRITE MODE whenever the Write Enable ($\overline{WE}$) and Chip Select ($\overline{CS}$) control inputs are in the low state. The status of the 8 output buffers during a write cycle is expalined below.

The WRITE cycle is initiated by the $\overline{WE}$ pulse going low provided that $\overline{CS}$ is also low. The leading edge of the $\overline{WE}$ pulse is used to latch the status of the address bus. $\overline{CS}$ if active (low) will also be latched. NOTE: $\overline{WE}$ is gated by $\overline{CS}$. If $\overline{CS}$ goes low after $\overline{WE}$, the Write Cycle will be initiated by $\overline{CS}$, and all timing will be referenced to that edge. $\overline{CS}$ and the Addresses will then be latched, and the cycle must be terminated by $\overline{WE}$ going high. The output bus if not already disabled will go to the high Z state t_{WEZ} after $\overline{WE}$. The latch signal, if at a logic high, will have no impact on the WRITE cycle. If latch is brought from a logic high to low prior to $\overline{WE}$ going active then the address inputs

and $\overline{CS}$ will be latched. NOTE: The Latch control ($\overline{L}$) will latch $\overline{CS}$ independent of the state, whereas $\overline{WE}$ will latch $\overline{CS}$ only when in the low state. Once latched, $\overline{CS}$ and the address inputs may be removed after the required hold times have been met.

Data in must be valid t_{DSW} prior to the low-to-high transition of $\overline{WE}$. The Data in lines must remain stable for t_{DHW} after $\overline{WE}$ goes inactive. The write control of the MK 4118 disables the data out buffers during the write cycle; however, output enable ($\overline{OE}$) should be used to disable the data out buffers to prevent bus contention between the input data and data that would be output upon completion of the write cycle.

READ/MODIFY/WRITE CYCLE

The MK4118 READ/MODIFY/WRITE cycle is merely a combination of the READ and WRITE cycle operations. The asynchronous or synchronous READ cycle may be combined with the WRITE operation. The status of DATA OUT bus will foll⋯ the operation outlined in the READ MODE or W

Figure 4.2(e)

Similarly the **write enable** pin $\overline{WE}$ corresponds to the **write** $\overline{WR}$ of a microprocessor, and allows data to be written into the RAM. However, there is no $\overline{RD}$ pin and it is assumed that whenever the device is not being written to, it is capable of being read. Data output is actually synchronised by the $\overline{OE}$, output enable signal.

(e) Pin 19 of this device is a special input known as a **latch** L. This pin can be used as the chip select when it is used with devices whose address bus may not be stable for the complete read cycle. The latch pin allows the address and the chip select data to be stored internally on the negative going edge of the L pulse and this in turn allows the data on the address or $\overline{CS}$ pins to be changed while a read cycle is in operation. This may be particularly useful in circuits where the address bus is multiplexed.

(f) It can be seen in the block diagram that the matrix of memory cells is arranged as $128 \times 8 \times 8$. Notice also that the address lines entering the Y decoder and the X decoder are not consecutive. This means that adjacent memory cells do not carry bits from adjacent data bytes. This is of no consequence to the system designer, since the bits are always returned from the appropriate storage location.

If this memory device were used in a Z80 microcomputer system it is most likely that the read and write cycle timing would correspond with the static read and write cycles given in the data sheet. The $\overline{OE}$ signal corresponds with the Z80 $\overline{RD}$ signal and the $\overline{WE}$, write signal corresponds with the Z80 $\overline{WR}$ signal. Chip Select $\overline{CS}$ would be provided as normal with the combination of the memory address and the $\overline{MREQ}$ control signal of the Z80.

4.2 DYNAMIC RAM

The amount of read/write memory in microprocessor based systems appears to be ever increasing. Microcomputer applications ranging from computerised games machines to minicomputers are very large memory users. They require RAM which is flexible in operation, tolerant of power supply noise, reliable, simple to interface and which offers the highest possible bit density. Dynamic RAMs are a natural choice for many applications.

They have a different construction from the static RAM in that each dynamic RAM cell is based on a single transistor and capacitor circuit. The capacitor stores a small charge which can be designated as a logic '1' when present or a '0' when absent. The chief problem with this arrangement is that the charge leaks away in a relatively short time and therefore the memory needs to be regularly refreshed. Typically, memory refreshing must take place every 2 ms. A previous chapter illustrated the circuit complexity required for typical dynamic RAM memory.

Almost every microcomputer with more than a few thousand bytes of RAM will use dynamic memory, and the most widely used chip has historically been a 16K $\times$ 1 device known as a MK4116.

MK4116 Dynamic Ram

The MK4116 represents a typical dynamic RAM although it has been superseded for many large memory circuits, its basic operation remains common to all memory sizes. *Figure 4.3* represents an extract from its data sheets, which show its main characteristics.

Main features

(a) Some of the most important information about the MK4116 is given on the first page of the data sheet. Unlike the static RAM, the dynamic RAM requires three power supplies +12 V, +5 V and −5 V, each of which are allowed a 10 per cent tolerance. This immediately causes a problem with systems where only +5 V is available and therefore dynamic RAMs tend not to be used in small systems. Even in larger computers, a system designer must provide three stable power supplies which requires additional expense and space.

(b) Power consumption of the device is very different between its active and standby

16,384 X 1-BIT DYNAMIC RAM
MK4116(P/N)-2/3

FEATURES

☐ Recognized industry standard 16-pin configuration from MOSTEK

☐ 150ns access time, 375ns cycle (MK 4116-2)
200ns access time, 375ns cycle (MK 4116-3)

☐ ± 10% tolerance on all power supplies (+12V, ±5V)

☐ Low power: 462mW active, 20mW standby (max)

☐ Output data controlled by $\overline{CAS}$ and unlatched at
end of cycle to allow two dimensional chip selection and extended page boundary

☐ Common I/O capability using ''early write''
operation

☐ Read-Modify-Write, $\overline{RAS}$-only refresh, and Page-
mode capability

☐ All inputs TTL compatible,low capacitance, and
protected against static charge

☐ 128 refresh cycles

☐ ECL compatible on VBB power supply (–5.7V)

DESCRIPTION

The MK 4116 is a new generation MOS dynamic
random access memory circuit organized as 16,384
words by 1 bit. As a state-of-the-art MOS memory
device, the MK 4116 (16K RAM) incorporates
advanced circuit techniques designed to provide
wide operating margins, both internally and to the
system user, while achieving performance levels
in speed and power previously seen only in MOSTEK's
high performance MK 4027 (4K RAM).

The technology used to fabricate the MK 4116 is
MOSTEK's double-poly, N-channel silicon gate,
POLY II ® process. This process, coupled with the
use of a single transistor dynamic storage cell, provides the maximum possible circuit density and
reliability, while maintaining high performance

capability. The use of dynamic circuitry throughout, including sense amplifiers, assures that power
dissipation is minimized without any sacrifice in
speed or operating margin. These factors combine
to make the MK 4116 a truly superior RAM product.

Multiplexed address inputs (a feature pioneered by
MOSTEK for its 4K RAMS) permits the MK 4116
to be packaged in a standard 16-pin DIP. This
recognized industry standard package configuration,
while compatible with widely available automated
testing and insertion equipment, provides highest
possible system bit densities and simplifies system
upgrade from 4K to 16K RAMs for new generation
applications. Non-critical clock timing requirements
allow use of the multiplexing technique while maintaining high performance.

FUNCTIONAL DIAGRAM

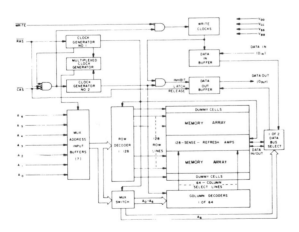

PIN CONNECTIONS

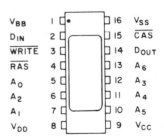

PIN NAMES

A_0-A_6	ADDRESS INPUTS
$\overline{CAS}$	COLUMN ADDRESS STROBE
D_{IN}	DATA IN
D_{OUT}	DATA OUT
$\overline{RAS}$	ROW ADDRESS STROBE
$\overline{WRITE}$	READ/WRITE INPUT
V_{BB}	POWER (–5V)
V_{CC}	POWER (+5V)
V_{DD}	POWER (+12V)
V_{SS}	GROUND

Figure 4.3(a)–(g) MK4116 data sheets (Courtesy Mostek)

ABSOLUTE MAXIMUM RATINGS*

Voltage on any pin relative to V$_{BB}$ . . . . −0.5V to +20V
Voltage on V$_{DD}$, V$_{CC}$ supplies relative to V$_{SS}$ . . . . −1.0V to +15.0V
V$_{BB}$−V$_{SS}$ (V$_{DD}$−V$_{SS}$ >0V) 0V
Operating temperature, T$_A$ (Ambient) . . . . 0°C to + 70°C
Storage temperature (Ambient) Ceramic . . . . −55°C to + 150°C
Storage temperature, (Ambient) Plastic . . . . −55°C to +125°C
Short circuit output current 50mA
Power dissipation . . . . 1 Watt

*Stresses greater than those listed under "Absolute Maximum Ratings" may cause permanent damage to the device. This is a stress rating only and functional operation of the device at these or any other conditions above those indicated in the operational sections of this specification is not implied. Exposure to absolute maximum rating conditions for extended periods may affect reliability.

RECOMMENDED DC OPERATING CONDITIONS[6]
(0°C ⩽ T$_A$ ⩽ 70°C)

PARAMETER	SYMBOL	MIN	TYP	MAX	UNITS	NOTES
Supply Voltage	V$_{DD}$	10.8	12.0	13.2	Volts	2
	V$_{CC}$	4.5	5.0	5.5	Volts	2,3
	V$_{SS}$	0	0	0	Volts	2
	V$_{BB}$	−4.5	−5.0	−5.7	Volts	2
Input High (Logic 1) Voltage, $\overline{RAS}$, $\overline{CAS}$, $\overline{WRITE}$	V$_{IHC}$	2.4	−	7.0	Volts	2
Input High (Logic 1) Voltage, all inputs except $\overline{RAS}$, $\overline{CAS}$ $\overline{WRITE}$	V$_{IH}$	2.2	−	7.0	Volts	2
Input Low (Logic 0) Voltage, all inputs	V$_{IL}$	−1.0	−	.8	Volts	2

DC ELECTRICAL CHARACTERISTICS
(0°C ⩽ TA ⩽ 70°C) (VDD = 12.0V ± 10%; VCC = 5.0V ±10%; −5.7V ⩽ VBB ⩽ −4.5V; VSS = 0V)

PARAMETER	SYMBOL	MIN	MAX	UNITS	NOTES
OPERATING CURRENT Average supply operating current ($\overline{RAS}$, $\overline{CAS}$ cycling; t$_{RC}$ = t$_{RC}$ Min	I$_{DD1}$ I$_{CC1}$ I$_{BB1}$		35 200	mA μA	4 5
STANDBY CURRENT Power supply standby current ($\overline{RAS}$ = V$_{IHC}$, D$_{OUT}$ = High Impedance)	I$_{DD2}$ I$_{CC2}$ I$_{BB2}$	−10	1.5 10 100	mA μA μA	
REFRESH CURRENT Average power supply current, refresh mode ($\overline{RAS}$ cycling, $\overline{CAS}$ = V$_{IHC}$; t$_{RC}$ = t$_{RC}$ Min	I$_{DD3}$ I$_{CC3}$ I$_{BB3}$	−10	25 10 200	mA μA μA	4
PAGE MODE CURRENT Average power supply current, page-mode operation ($\overline{RAS}$ = V$_{IL}$, $\overline{CAS}$ cycling; t$_{PC}$ = t$_{PC}$ Min	I$_{DD4}$ I$_{CC4}$ I$_{BB4}$		27 200	mA μA	4 5
INPUT LEAKAGE Input leakage current, any input (V$_{BB}$ = −5V, 0V ⩽ V$_{IN}$ ⩽ +7.0V, all other pins not under test = 0 volts)	I$_{I(L)}$	−10	10	μA	
OUTPUT LEAKAGE Output leakage current (D$_{OUT}$ is disabled, 0V ⩽ V$_{OUT}$ ⩽ +5.5V)	I$_{0(L)}$	−10	10	μA	
OUTPUT LEVELS Output high (Logic 1) voltage (I$_{OUT}$ = −5mA)	V$_{OH}$	2.4		Volts	3
Output low (Logic 0) voltage (I$_{OUT}$ = 4.2 mA)	V$_{OL}$		0.4	Volts	

NOTES:

1.   T$_A$ is specified here for operation at frequencies to t$_{RC}$ ⩾ t$_{RC}$ (min). Operation at higher cycle rates with reduced ambient temperatures and higher power dissipation is permissible, however, provided AC operating parameters are met. See figure 1 for derating curve.

2.   All voltages referenced to V$_{SS}$.

3.   Output voltage will swing from V$_{SS}$ to V$_{CC}$ when activated with no current loading. For purposes of maintaining data in standby

mode, V$_{CC}$ may be reduced to V$_{SS}$ without affecting refresh operations or data retention. However, the V$_{OH}$ (min) specification is not guaranteed in this mode.

4.   I$_{DD1}$, I$_{DD3}$, and I$_{DD4}$ depend on cycle rate. See figures 2,3, and 4 for I$_{DD}$ limits at other cycle rates.

5.   I$_{CC1}$ and I$_{CC4}$ depend upon output loading. During readout of high level data V$_{CC}$ is connected through a low impedance (135 Ω typ) to data out. At all other times I$_{CC}$ consists of leakage currents only.

Figure 4.3(b)

ELECTRICAL CHARACTERISTICS AND RECOMMENDED AC OPERATING CONDITIONS (6,7,8)

$(0°C \leq T_A \leq 70°C)^1$ (V_{DD} = 12.0V ± 10%; V_{CC} = 5.0V ±10%, V_{SS} = 0V, V_{BB} = **-5.7V** $\leq$ **VBB** $\leq$ **-4.5V**)

PARAMETER	SYMBOL	MK 4116-2 MIN	MK 4116-2 MAX	MK 4116-3 MIN	MK 4116-3 MAX	UNITS	NOTES
Random read or write cycle time	tRC	375		375		ns	9
Read-write cycle time	tRWC	375		375		ns	9
Read modify write cycle time	tRMW	320		405		ns	9
Page mode cycle time	tPC	170		225		ns	9
Access time from $\overline{RAS}$	tRAC		150		200	ns	10,12
Access time from $\overline{CAS}$	tCAC		100		135	ns	11,12
Output buffer turn-off delay	tOFF	0	40	0	50	ns	13
Transition time (rise and fall)	tT	3	35	3	50	ns	8
$\overline{RAS}$ precharge time	tRP	100		120		ns	
$\overline{RAS}$ pulse width	tRAS	150	10,000	200	10,000	ns	
$\overline{RAS}$ hold time	tRSH	100		135		ns	
$\overline{CAS}$ hold time	tCSH	150		200		ns	
$\overline{CAS}$ pulse width	tCAS	100	10,000	135	10,000	ns	
$\overline{RAS}$ to $\overline{CAS}$ delay time	tRCD	20	50	25	65	ns	14
$\overline{CAS}$ to $\overline{RAS}$ precharge time	tCRP	−20		−20		ns	
Row Address set-up time	tASR	0		0		ns	
Row Address hold time	tRAH	20		25		ns	
Column Address set-up time	tASC	−10		−10		ns	
Column Address hold time	tCAH	45		55		ns	
Column Address hold time referenced to $\overline{RAS}$	tAR	95		120		ns	
Read command set-up time	tRCS	0		0		ns	
Read command hold time	tRCH	0		0		ns	
Write command hold time	tWCH	45		55		ns	
Write command hold time referenced to $\overline{RAS}$	tWCR	95		120		ns	
Write command pulse width	tWP	45		55		ns	
Write command to $\overline{RAS}$ lead time	tRWL	50		70		ns	
Write command to $\overline{CAS}$ lead time	tCWL	50		70		ns	
Data-in set-up time	tDS	0		0		ns	15
Data-in hold time	tDH	45		55		ns	15
Data-in hold time referenced to $\overline{RAS}$	tDHR	95		120		ns	
$\overline{CAS}$ precharge time (for page-mode cycle only)	tCP	60		80		ns	
Refresh period	tREF		2		2	ms	
$\overline{WRITE}$ command set-up time	tWCS	−20		−20		ns	16
$\overline{CAS}$ to $\overline{WRITE}$ delay	tCWD	60		80		ns	16
$\overline{RAS}$ to $\overline{WRITE}$ delay	tRWD	110		145		ns	16

NOTES (Continued)

6. Several cycles are required after power-up before proper device operation is achieved. Any 8 cycles which perform refresh are adequate for this purpose.
7. AC measurements assume tT = 5ns.
8. VIHC (min) or VIH (min) and VIL (max) are reference levels for measuring timing of input signals. Also transition times are measured between VIHC or VIH and VIL.
9. The specifications for tRC (min) tRMW (min) and tRWC (min) are used only to indicate cycle time at which proper operation over the full temperature range (0°C ≤ TA ≤ 70°C) is assured
10. Assumes that tRCD ≤ tRCD (Max). If tRCD is greater than the maximum recommended value shown in this table, tRAC will increase by the amount that tRCD exceeds the value shown.
11. Assumes that tRCD (max).
12. Measured with a load equivalent to 2 TTL loads and 100pF.
13. tOFF (max) defines the time at which the output achieves the open circuit condition and is not referenced to output voltage levels.

14. Operation within the tRCD (max) limit insures that tRAC (max) can be met. tRCD (max) is specified as a reference point only if tRCD is greater than the specified tRCD (max) limit, then access time is controlled exclusively by tCAC.
15. These parameters are referenced to CAS leading edge in early write cycles and to WRITE leading edge in delayed write or read-modify-write cycles.
16. tWCS, tCWD and tRWD are restrictive operating parameters in read write and read modify write cycles only. If tWCS ≥ tWCS (min), the cycle is an early write cycle and the data out pin will remain open circuit (high impedance) throughout the entire cycle; If tCWD ≥ tCWD (min) and tRWD ≥ tRWD (min), the cycle is a read-write cycle and the data out will contain data read from the selected cell. If neither of the above sets of conditions is satisfied the condition of the data out (at access time) is indeterminate.
17. Effective capacitance calculated from the equation C = $\frac{1 \Delta t}{\Delta V}$ ΔV with Δ = 3 volts and power supplies at nominal levels
18. CAS = VIHC to disable DOUT.

Figure 4.3(c)

READ CYCLE

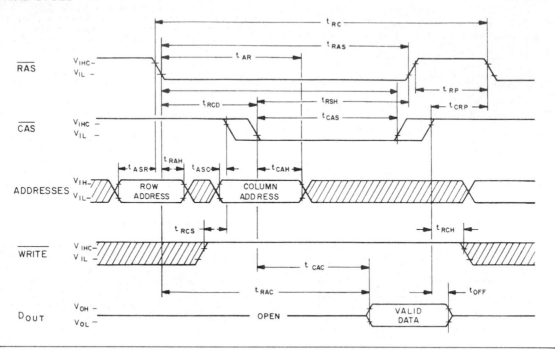

WRITE CYCLE (EARLY WRITE)

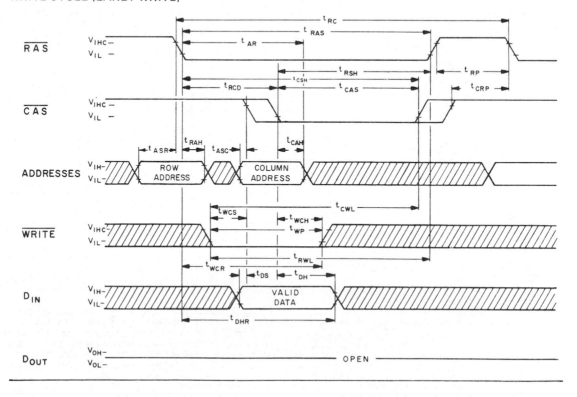

Figure 4.3(d)

DESCRIPTION (continued)

System oriented features include ± 10% tolerance on all power supplies, direct interfacing capability with high performance logic families such as Schottky TTL, maximum input noise immunity to minimize "false triggering" of the inputs (a common cause of soft errors), on-chip address and data registers which eliminate the need for interface registers, and two chip select methods to allow the user to determine the appropriate speed/power characteristics of his memory system. The MK 4116 also incorporates several flexible timing/operating modes. In addition to the usual read, write, and read-modify-write cycles, the MK 4116 is capable of delayed write cycles, page-mode operation and $\overline{RAS}$-only refresh. Proper control of the clock inputs($\overline{RAS}$, $\overline{CAS}$ and $\overline{WRITE}$) allows common I/O capability, two dimensional chip selection, and extended page boundaries (when operating in page mode).

ADDRESSING

The 14 address bits required to decode 1 of the 16,384 cell locations within the MK 4116 are multiplexed onto the 7 address inputs and latched into the on-chip address latches by externally applying two negative going TTL-level clocks. The first clock, the Row Address Strobe ($\overline{RAS}$), latches the 7 row address bits into the chip. The second clock, the Column Address Strobe ($\overline{CAS}$), subsequently latches the 7 column address bits into the chip. Each of these signals, $\overline{RAS}$ and $\overline{CAS}$, triggers a sequence of events which are controlled by different delayed internal clocks. The two clock chains are linked together logically in such a way that the address multiplexing operation is done outside of the critical path timing sequence for read data access. The later events in the $\overline{CAS}$ clock sequence are inhibited until the occurence of a delayed signal derived from the $\overline{RAS}$ clock chain. This "gated $\overline{CAS}$" feature allows the $\overline{CAS}$ clock to be externally activated as soon as the Row Address Hold Time specification (t_{RAH}) has been satisfied and the address inputs have been changed from Row address to Column address information.

Note that $\overline{CAS}$ can be activated at any time after t_{RAH} and it will have no effect on the worst case data access time (t_{RAC}) up to the point in time when the delayed row clock no longer inhibits the remaining sequence of column clocks. Two timing endpoints result from the internal gating of $\overline{CAS}$ which are called t_{RCD} (min) and t_{RCD} (max). No data storage or reading errors will result if $\overline{CAS}$ is applied to the MK 4116 at a point in time beyond the t_{RCD} (max) limit. However, access time will then be determined exclusively by the access time from $\overline{CAS}$ (t_{CAC}) rather than from $\overline{RAS}$ (t_{RAC}), and access time from $\overline{RAS}$ will be lengthened by the amount that t_{RCD} exceeds the t_{RCD} (max) limit.

DATA INPUT/OUTPUT

Data to be written into a selected cell is latched into an on-chip register by a combination of $\overline{WRITE}$ and $\overline{CAS}$ while $\overline{RAS}$ is active. The later of the signals ($\overline{WRITE}$ or $\overline{CAS}$) to make its negative transition is the strobe for the Data In (D_{IN}) register. This permits several options in the write cycle timing. In a write cycle, if the $\overline{WRITE}$ input is brought low (active)

prior to $\overline{CAS}$, the D_{IN} is strobed by $\overline{CAS}$, and the set-up and hold times are referenced to $\overline{CAS}$. If the input data is not available at $\overline{CAS}$ time or if it is desired that the cycle be a read-write cycle, the $\overline{WRITE}$ signal will be delayed until after $\overline{CAS}$ has made its negative transition. In this "delayed write cycle" the data input set-up and hold times are referenced to the negative edge of $\overline{WRITE}$ rather than $\overline{CAS}$. (To illustrate this feature, D_{IN} is referenced to $\overline{WRITE}$ in the timing diagrams depicting the read-write and page-mode write cycles while the "early write" cycle diagram shows D_{IN} referenced to $\overline{CAS}$).

Data is retrieved from the memory in a read cycle by maintaining $\overline{WRITE}$ in the inactive or high state throughout the portion of the memory cycle in which $\overline{CAS}$ is active (low). Data read from the selected cell will be available at the output within the specified access time.

DATA OUTPUT CONTROL

The normal condition of the Data Output (D_{OUT}) of the MK 4116 is the high impedance (open-circuit) state. That is to say, anytime $\overline{CAS}$ is at a high level, the D_{OUT} pin will be floating. The only time the output will turn on and contain either a logic 0 or logic 1 is at access time during a read cycle. D_{OUT} will remain valid from access time until $\overline{CAS}$ is taken back to the inactive (high level) condition.

If the memory cycle in progress is a read, read-modify write, or a delayed write cycle, then the data output will go from the high impedance state to the active condition, and at access time will contain the data read from the selected cell. This output data is the same polarity (not inverted) as the input data. Once having gone active, the output will remain valid until $\overline{CAS}$ is taken to the precharge (logic 1) state, whether or not $\overline{RAS}$ goes into precharge.

If the cycle in progress is an "early-write" cycle ($\overline{WRITE}$ active before $\overline{CAS}$ goes active), then the output pin will maintain the high impedance state throughout the entire cycle. Note that with this type of output configuration, the user is given full control of the D_{OUT} pin simply by controlling the placement of $\overline{WRITE}$ command during a write cycle, and the pulse width of the Column Address Strobe during read operations. Note also that even though data is not latched at the output, data can remain valid from access time until the beginning of a subsequent cycle without paying any penalty in overall memory cycle time (stretching the cycle).

This type of output operation results in some very significant system implications.

Common I/O Operation — If all write operations are handled in the "early write" mode, then D_{IN} can be connected directly to D_{OUT} for a common I/O data bus.

Data Output Control — D_{OUT} will remain valid during a read cycle from t_{CAC} until $\overline{CAS}$ goes back to a high level (precharge), allowing data to be valid from one cycle up until a new memory cycle begins with no penalty in cycle time. This also makes the $\overline{RAS}/\overline{CAS}$ clock timing relationship very flexible.

Two Methods of Chip Selection — Since D_{OUT}

Figure 4.3(e)

is not latched, $\overline{CAS}$ is not required to turn off the outputs of unselected memory devices in a matrix. This means that both $\overline{CAS}$ and/or $\overline{RAS}$ can be decoded for chip selection. If both $\overline{RAS}$ and $\overline{CAS}$ are decoded, then a two dimensional (X,Y) chip select array can be realized.

Extended Page Boundary — Page-mode operation allows for successive memory cycles at multiple column locations of the same row address. By decoding $\overline{CAS}$ as a page cycle select signal, the page boundary can be extended beyond the 128 column locations in a single chip. (See page-mode operation).

OUTPUT INTERFACE CHARACTERISTICS

The three state data output buffer presents the data output pin with a low impedance to V_{CC} for a logic 1 and a low impedance to V_{SS} for a logic 0. The effective resistance to V_{CC} (logic 1 state) is 420 Ω maximum and 135Ω typically. The resistance to V_{SS} (logic 0 state) is 95 Ω maximum and 35 Ω typically. The separate V_{CC} pin allows the output buffer to be powered from the supply voltage of the logic to which the chip is interfaced. During battery standby operation, the V_{CC} pin may have power removed without affecting the MK 4116 refresh operation. This allows all system logic except the $\overline{RAS}$ timing circuitry and the refresh address logic to be turned off during battery standby to conserve power.

PAGE MODE OPERATION

The "Page Mode" feature of the MK 4116 allows for successive memory operations at multiple column locations of the same row address with increased speed without an increase in power. This is done by strobing the row address into the chip and maintaining the $\overline{RAS}$ signal at a logic 0 throughout all successive memory cycles in which the row address is common. This "page-mode" of operation will not dissipate the power associated with the negative going edge of $\overline{RAS}$. Also, the time required for strobing in a new row address is eliminated, thereby decreasing the access and cycle times.

The page boundary of a single MK 4116 is limited to the 128 column locations determined by all combinations of the 7 column address bits. However, in system applications which utilize more than 16,384 data words, (more than one 16K memory block), the page boundary can be extended by using $\overline{CAS}$ rather than $\overline{RAS}$ as the chip select signal. $\overline{RAS}$ is applied to all devices to latch the row address into each device and then $\overline{CAS}$ is decoded and serves as a page cycle select signal. Only those devices which receive both $\overline{RAS}$ and $\overline{CAS}$ signals will execute a read or write cycle.

REFRESH

Refresh of the dynamic cell matrix is accomplished by performing a memory cycle at each of the 128 row addresses within each 2 millisecond time interval. Although any normal memory cycle will perform the refresh operation, this function is most easily accomplished with "$\overline{RAS}$-only" cycles. $\overline{RAS}$-only refresh results in a substantial reduction in operating power. This reduction in power is reflected in the I_{DD3} specification.

POWER CONSIDERATIONS

Most of the circuitry used in the MK 4116 is dynamic and most of the power drawn is the result of an address strobe edge. Consequently, the dynamic power is primarily a function of operating frequency rather than active duty cycle (refer to the MK 4116 current waveforms in figure 5). This current characteristic of the MK 4116 precludes inadvertent burn out of the device in the event that the clock inputs become shorted to ground due to system malfunction.

Although no particular power supply noise restriction exists other than the supply voltages remain within the specified tolerance limits, adequate decoupling should be provided to suppress high frequency noise resulting from the transient current of the device. This insures optimum system performance and reliability. Bulk capacitance requirements are minimal since the MK 4116 draws very little steady state (DC) current.

In system applications requiring lower power dissipation ,the operating frequency (cycle rate) of the MK 4116 can be reduced and the (guaranteed maximum) average power dissipation of the device will be lowered in accordance with the I_{DD1} (max) spec limit curve illustrated in figure 2 . NOTE: The MK 4116 family is guaranteed to have a maximum I_{DD1} requirement of 35mA @ 375ns cycle (320ns cycle for the –2) with an ambient temperature range from 0° to 70°C. A lower operating frequency, for example 1 microsecond cycle, results in a reduced maximum Idd1 requirement of under 20mA with an ambient temperature range from 0° to 70°C.

It is possible the MK4116 family (–2 and 3 speed selections for example) at frequencies higher than specified, provided all AC operating parameters are met. Operation at shorter cycle times (<tRC min) results in higher power dissipation and, therefore, a reduction in ambient temperature is required. Refer to Figure 1 for derating curve.

NOTE: Additional power supply tolerance has been included on the VBB supply to allow direct interface capability with both –5V systems –5.2V ECL systems.

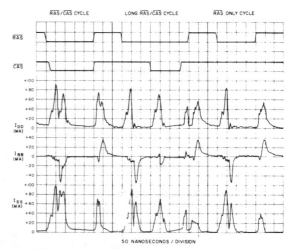

Fig. 5 Typical Current Waveforms

Figure 4.3(f)

Although $\overline{RAS}$ and/or $\overline{CAS}$ can be decoded and used as a chip select signal for the MK 4116, overall system power is minimized if the Row Address Strobe ($\overline{RAS}$) is used for this purpose. All unselected devices (those which do not receive a $\overline{RAS}$) will remain in a low power (standby) mode regardless of the state of $\overline{CAS}$.

POWER UP

The MK 4116 requires no particular power supply sequencing so long as the Absolute Maximum Rating Conditions are observed. However, in order to insure compliance with the Absolute Maximum Ratings, MOSTEK recommends sequencing of power supplies such that V_{BB} is applied first and removed last. V_{BB} should never be more positive than V_{SS} when power is applied to V_{DD}.

Under system failure conditions in which one or more supplies exceed the specified limits significant additional margin against catastrophic device failure may be achieved by forcing $\overline{RAS}$ and $\overline{CAS}$ to the inactive state (high level).

After power is applied to the device, the MK 4116 requires several cycles before proper device operation is achieved. Any 8 cycles which perform refresh are adequate for this purpose.

Figure 4.3(g)

modes, 426 mW compared with 20 mW. This means that a lot of power can be saved because for most of the time the memory devices will be on standby.

(c) The device is packaged in a standard 16 pin dual in line pack. The most obvious feature of this is the number of address lines, only 7, which mean that the address bus must be multiplexed onto these pins from the external circuit.

(d) The internal structure of the memory matrix is much more complex than a static RAM. Notice that it is a matrix of 128 x 64 memory cells with dummy cells at either side and refresh amplifiers between two halves of the matrix.

(e) The second page of the characteristics (*Figure 4.3(b)*, page 92) shows the operating conditions and here it is possible to see, for example, that the maximum operating current is 35 mA while the standby current is 1.5 mA.

Most of the current is drawn from the I_{DD} supply, i.e. $+12$ V; and in comparison, the $+5$ and -5 V supplies only supply micro amps of current.

(f) The timing chart on the next page (*Figure 4.3(c)*, page 93) shows a number of important timings relating to diagrams that follow it. One significant time is the refresh period on the fourth line from the bottom. This shows that the refresh time has a maximum value of 2 ms. This means that if the memory is not refreshed within this time all its data will be lost.

(g) The timing diagrams (*Figure 4.3(d)*, page 94) clearly show the operation of the memory. For example, the read cycle shows how the row address must be valid when the $\overline{RAS}$ signal goes to its low state, and the column address must be valid when the $\overline{CAS}$ signal goes low. Data can be read out of the memory after a time t_{rac}, which by reference to the timing chart has a maximum value of 200 ns for the MK4116–3. As soon as the $\overline{CAS}$ signal rises to its logic '1' state, the data becomes invalid after a short time period, and the read cycle is complete.

Similarly the write cycle requires both the row and column addresses to be supplied at the appropriate moment and these are synchronised with the $\overline{RAS}$ and $\overline{CAS}$ signals. In the diagram shown the falling edge of the $\overline{CS}$ signal also strobes the data into memory, as long as it is held for a minimum period of t_{dh}, which is 55 ns in the MK4116–3.

4.3 READ ONLY MEMORY

Read Only Memory is a vital part of every computer system. It is the type of memory that can only be read, but it will retain its data even when the power is removed. Therefore it can be used to hold programs that may operate as soon as the system is switched on. In mass-produced systems, the programs in ROM may be built in at the manufacturing stage, or they may be programmed by the equipment manufacturer.

64K-BIT READ-ONLY MEMORY
MK36000(P/N)-4/5

FEATURES

☐ MK36000 8K x 8 Organization—
   "Edge Activated" * operation ($\overline{CE}$)

☐ Access Time/Cycle Time

P/N	Access	Cycle
MK36000-4	250ns	375ns
MK36000-5	300ns	450ns

☐ Single +5V ± 10% Power Supply

☐ Standard 24 pin DIP (EPROM Pin Out Compatible)

☐ Low Power Dissipation - 220mW Max Active

☐ Low Standby Power Dissipation—35mW Max.
   ($\overline{CE}$ High)

☐ On chip latches for addresses

☐ Inputs and three-state outputs-TTL compatible

☐ Outputs drive 2 TTL loads and 100 pF

DESCRIPTION

The MK36000 is a new generation N-channel silicon gate MOS Read Only Memory, organized as 8192 words by 8 bits. As a state-of-the-art device, the MK 36000 incorporates advanced circuit techniques designed to provide maximum circuit density and reliability with the highest possible performance, while maintaining low power dissipation and wide operating margins.

The MK36000 utilizes what is fast becoming an industry standard method of device operation. Use of a static storage cell with clocked control periphery allows the circuit to be put into an automatic low power standby mode. This is accomplished by maintaining the chip enable ($\overline{CE}$) input at a TTL high level. In this mode, power dissipation is reduced to typically 35mW, as compared to unclocked devices which

draw full power continuously. In system operation, a device is selected by the $\overline{CE}$ input, while all others are in a low power mode, reducing the overall system power. Lower power means reduced power supply cost, less heat to dissipate and an increase in device and system reliability.

The edge activated chip enable also means greater system flexibility and an increase in system speed. The MK36000 features onboard address latches controlled by the $\overline{CE}$ input. Once the address hold time specification has been met, new address data can be applied in anticipation of the next cycle. Outputs can be wire-'OR'ed together, and a specific device can be selected by utilizing the $\overline{CE}$ input with no bus conflict on the outputs. The $\overline{CE}$ input allows the fastest access times yet available in 5 volt only

FUNCTIONAL DIAGRAM (MK 36000)   PIN CONNECTIONS

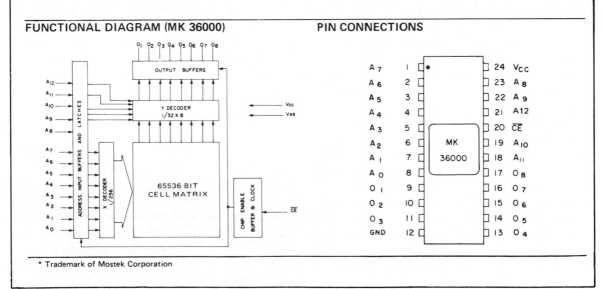

* Trademark of Mostek Corporation

Figure 4.4(a)—(d)   MK36000 MOS read only memory (Courtesy Mostek)

ABSOLUTE MAXIMUM RATINGS*

Voltage on Any Terminal Relative to V_{SS} . . . . −1.0 V to + 7V

Operating Temperature T_A (Ambient) . . . . 0°C to + 70°C

Storage Temperature — Ceramic (Ambient) . . . . −65°C to + 150°C

Storage Temperature — Plastic (Ambient) . . . . −55°C to + 125°C

Power Dissipation . . . . 1 Watt

*Stresses greater than those listed under "Absolute Maximum Ratings" may cause permanent damage to the device. This is a stress rating only and functional operation of the device at these or any other conditions above those indicated in the operating sections of this specification is not implied. Exposure to absolute maximum rating conditions for extended periods may affect device reliability.

RECOMMENDED DC OPERATING CONDITIONS[6]
(0°C ⩽ T_A ⩽ + 70°C)

	PARAMETER	MIN	TYP	MAX	UNITS	NOTES
V_{CC}	Power Supply Voltage	4.5	5.0	5.5	Volts	6
V_{IL}	Input Logic 0 Voltage	−1.0		0.8	Volts	
V_{IH}	Input Logic 1 Voltage	2.0		V_{CC}	Volts	

D C ELECTRICAL CHARACTERISTICS
(V_{CC} = 5V ± 10%) (0°C ⩽ T_A ⩽ + 70°C)[6]

	PARAMETER	MIN	TYP	MAX	UNITS	NOTES
I_{CC1}	V_{CC} Power Supply Current (Active)			40	mA	1
I_{CC2}	V_{CC} Power Supply Current (Standby)			8	mA	7
$I_{I(L)}$	Input Leakage Current	−10		10	μA	2
$I_{O(L)}$	Output Leakage Current	−10		10	μA	3
V_{OL}	Output Logic "0" Voltage @ I_{OUT} = 3.3mA			0.4	volts	
V_{OH}	Output Logic "1" Voltage @ I_{OUT} = −220 μA	2.4			volts	

A C ELECTRICAL CHARACTERISTICS
(V_{CC} = 5V ± 10%) (0°C ⩽ T_A ⩽ + 70°C)[6]

	PARAMETER	-4 MIN	-4 MAX	-5 MIN	-5 MAX	UNITS	NOTES
t_C	Cycle Time	375		450		ns	4
t_{CE}	$\overline{CE}$ Pulse Width	250		300			4
t_{AC}	$\overline{CE}$ Access Time		250		300	ns	4
t_{OFF}	Output Turn Off Delay		60		75	ns	4
t_{AH}	Address Hold Time Referenced to $\overline{CE}$	60		75		ns	
t_{AS}	Address Setup Time Referenced to $\overline{CE}$	0		0		ns	
t_P	$\overline{CE}$ Precharge Time	125		150		ns	

NOTES:

1. Current is proportional to cycle rate. I_{CC1} is measured at the specified minimum cycle time.
2. V_{IN} = 0V to 5.5V (V_{CC} = 5V)
3. Device unselected; V_{OUT} = 0V to 5.5V
4. Measured with 2 TTL loads and 100pF, transistion times = 20ns
5. Capacitance measured with Boonton Meter or effective capacitance calculated from the equation:

$$C = \frac{\Delta Q}{\Delta V} \text{ with } \Delta V = 3 \text{ volts}$$

6. A minimum 100 μs time delay is required after the application of V_{CC} (+5) before proper device operation is achieved. $\overline{CE}$ must be at VIH for this time period.

7. $\overline{CE}$ high.

Figure 4.4(b)

CAPACITANCE
(0°C ≤ TA ≤ 70°C)

	PARAMETER	TYP	MAX	UNITS	NOTES
CI	Input Capacitance	5	8	pF	5
CO	Output Capacitance	7	15	pF	5

TIMING DIAGRAM

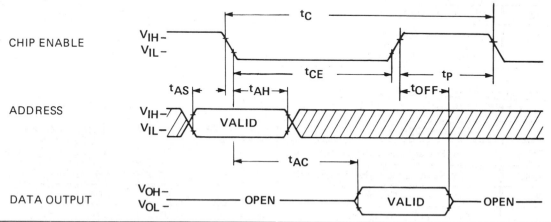

MK36000 ROM PUNCHED CARD CODING FORMAT (1 & 6)

FIRST CARD

COLS	INFORMATION FIELD
1-30	Customer
31-50	Customer Part Number
60-72	MOSTEK Part Number (2)

SECOND CARD

COLS	INFORMATION FIELD
1-30	Engineer at Customer Site
31-50	Direct Phone Number for Engineer

THIRD CARD

COLS	INFORMATION FIELD
1-5	MOSTEK Part Number (2)

FOURTH CARD

COLS	INFORMATION FIELD
1-9	Data Format (3)
15-28	Logic - ("Positive Logic" or "Negative Logic")
35-57	Verification Code (4)

DATA FORMAT

512 data cards (16 data words/card) with the following format:

COLS	INFORMATION FIELD
1-4	Four digit octal address of first output word on card
5-7	Three digit octal output word specified by address in column 1-4
8-52	Next fifteen output words, each word consists of three octal digits.

NOTES:

1. Positive or negative logic formats are accepted as noted in the fourth card.

2. Assigned by MOSTEK; may be left blank.

3. MOSTEK punched card coding format should be used. Punch "MOSTEK" starting in column one.

4. Punches as: (a) VERIFICATION HOLD - i.e. customer verification of the data as reproduced by MOSTEK is required prior to production of the ROM. To accomplish this MOSTEK supplies a copy of its Customer Verification Data Sheet (CVDS) to the customer.

   (b) VERIFICATION PROCESS - i.e. the customer will receive a CVDS but production will begin prior to receipt of customer verification; (c) VERIFICATION NOT NEEDED - i.e. the customer will not receive a CVDS and production will begin immediately.

5. 512 cards for MK 36000

6. Please consult MOSTEK ROM Programming Guide for further details on other formats.

Figure 4.4(c)

DESCRIPTION (Continued)

ROM's and imposes no loss in system operating flexibility over an unclocked device.

Other system oriented features include fully TTL compatible inputs and outputs. The three state outputs, controlled by the $\overline{CE}$ input, will drive a minimum of 2 standard TTL loads. The MK36000 operates from a single +5 volt power supply with a wide ± 10% tolerance, providing the widest operating margins available. The MK36000 is packaged in the industry standard 24 pin DIP.

Any application requiring a high performance, high bit density ROM can be satisfied by the MK36000 ROM. This device is ideally suited for 8 bit microprocessor systems such as those which utilize the Z-80. It can offer significant cost advantages over PROM.

OPERATION

The MK36000 is controlled by the chip enable ($\overline{CE}$) input. A negative going edge at the $\overline{CE}$ input will activate the device as well as strobe and latch the inputs into the onchip address registers. At access time the outputs will become active and contain the data read from the selected location. The outputs will remain latched and active until $\overline{CE}$ is returned to the inactive state.

Programming Data

MOSTEK is now able to utilize a wide spectrum of data input formats and media. Those presently available are listed in the following table:

Table 1

Acceptable Media	Acceptable Format
CARDS	MOSTEK
PAPER TAPE	INTEL CARD
PROMS	INTEL TAPE
DATA LINK	EA
	MOSTEK F-8
	MOTOROLA 6800

Figure 4.4(d)

In large-scale production the device used is generally a MOS ROM. For small-scale production an EPROM would normally be used.

MOS Read Only Memory

The MK36000 is an example of a typical MOS Read Only Memory. It is fully TTL compatible and comes in a number of speeds to suit the needs of most system designers.

This device has a number of features that are common to ROMs.

Main features

(a) The MK36000 is organised as an 8K × 8 device and has a 24-pin package. This package is deliberately chosen to be compatible with other 'bytewide' devices by using standard pins for the bus signals. This allows compatibility between the EPROM used to prototype any device and the final ROM used at the mass production stage.

(b) The power supply requirements are very straightforward – only a single +5 V rail. Power dissipation varies between 220 mW in its active state to 35 mW in its standby mode.

(c) The access time of the chip is relatively slow, between 250 and 300 ns. This is typical of large ROMs, but it makes them more difficult to interface with fast microprocessors, since occasionally a **wait** state must be added to the microprocessor machine cycle while the data from the memory is arriving. The memory cycle time (the time that is required between one read operation and the next) is slightly longer than the access time, as can be seen in the timing diagram (*Figure 4.4(c)*). This shows the one mode of operation of which the device is capable, i.e. reading.

(d) One interesting feature of the device data sheet is that it also gives an indication of the punched cards that must be supplied to the manufacturer in order to have a device programmed. These punched cards could be read directly by a machine which would allow the program to be implemented without any direct human intervention. The first four cards contain some basic information about the ROM to be produced, and then the main program is held on the following 512 data cards as indicated. Each data card holds 16 data words and the information must be specified in octal format. The last paragraph of the data sheet indicates that the programming data may also be supplied in different formats and this would be more normal for modern PROM devices. For example, data could be supplied on cards, paper tape, other PROMs, via data link, etc. Data for most modern PROMs could also be supplied on floppy disk in a specified file format.

EPROM

The **EPROM** is the most widely used Read Only Memory for prototype and small-scale production devices. They have the very useful characteristic that they can be programmed relatively easily and that the program may be erased at will should it require changing for any reason.

The programming requirements of each EPROM are slightly different, but with the range of EPROM programmers available, all EPROM types can be covered relatively easy by changes to the software used in the programmer.

Programming an EPROM typically involves the following steps.

(a) Selecting the appropriate EPROM type on the programmer either from a switch or keyboard.
(b) Inserting the EPROM.
(c) Preparing the EPROM programmer to receive data from the keyboard, a tape or more generally another computer. Most EPROM programmers accept .HEX files which are created during the Assembly process.
(d) Transferring the program to the EPROM programmer.
(e) Pressing the 'Program' button.

Most modern programmers check that the target EPROM is capable of receiving the program and give appropriate error messages if it is not.

Erasing an EPROM can be achieved by exposure to an ultraviolet light of the appropriate wavelength and intensity, and the data sheet for the typical EPROM chosen, the MK2716, indicates the erasing requirements.

MK2716 2048 × 8 EPROM

As with most of the other memory devices considered so far, the important features of the MK2716 are given on the first page of its data sheet (*Figure 4.5(a)*).

The 2716 has a number of interesting features which make it a very popular EPROM.

Main features

(a) The 2716 is pin compatible with some of the previous devices such as the MK36000 and this makes it possible to develop a program in the 2716 EPROM for downloading when complete into a mask programmable ROM.
(b) In common with other devices this has a single +5 V power supply requirement. However, the MODE selection box indicates that +25 V is required to program the device.
(c) The access time is relatively slow in comparison with most RAM chips and the fastest time given is 350 ns. This slow time means that when an EPROM is used in the circuit there is occasionally a need to add **wait** states to the microprocessor cycle time.
(d) The timing diagram of the read cycle (*Figure 4.5(c)*) shows that the device behaves in a very conventional way when connected to a computer system. Once the address is set up on the address lines and the output enable line is taken low, data appears on the output pins after a short time. The total time which this takes, t_{ACC}, is the ACCESS time of the device and is between 350 and 450 ns.
(e) Of considerable interest in this case, is the programming method for the device. This is shown in the program mode timing diagram (*Figure 4.5(e)*).

Programming may be done to either an individual address within the EPROM, or sequentially through the range of addresses. Normally, an EPROM programmer would perform this function and its operation would be controlled by built-in software.

The MK2716 is supplied completely erased, with all bits at a logic '1' level. To program the device the V_{pp} pin must be held at +25 V, and the OE pin at +5 V. When the program pin is pulsed from low to high and back to low again, the data present on the data pins is programmed into the address which must be present on the address pins. The A.C. characteristics chart indicates that the minimum

2048 x 8 BIT PROM
Electrically Programmable/Ultraviolet Erasable ROM
MK2716 (T)-6/7/8

FEATURES

☐ Replacement for popular 2048 x 8 bit 2716 type EPROM

☐ Single +5 volt power supply during READ operation

☐ Fast Access Time in READ mode

P/N	Access Time
MK2716-6	350ns
MK2716-7	390ns
MK2716-8	450ns

☐ Low Power Dissipation: 525 mW max active

☐ Power Down mode: 132 mW max standby

☐ Three State Output OR-tie capability

☐ Five modes of operation for greater system flexibility (see Table)

☐ Single programming requirement: single location programming with one 50 msec pulse

☐ Pin Compatible with MK34000 16K ROM

☐ TTL compatible in all operating modes

☐ Standard 24 pin DIP with transparent lid

DESCRIPTION

The MK2716 is a 2048x8 bit electrically programmable/ultraviolet erasable Read Only Memory. The circuit is fabricated with MOSTEK's advanced N-channel silicon gate technology for the highest performance and reliability. The MK2716 offers significant advances over hardwired logic in cost, system flexibility, turnaround time and performance.

The MK2716 has many useful system oriented features including a STANDBY mode of operation which lowers the device power from 525 mW maximum active power to 132 mW maximum for an overall savings of 75%.

MODE SELECTION

MODE \ PIN	$\overline{CE}$/PGM (18)	$\overline{OE}$ (20)	VPP (21)	OUTPUTS
READ	VIL	VIL	+5	Valid Out
STANDBY	VIH	Don't Care	+5	Open
PROGRAM	Pulsed VIL to VIH	VIH	+25	Input
PROGRAM VERIFY	VIL	VIL	+25	Valid Out
PROGRAM INHIBIT	VIL	VIH	+25	Open
VCC(24) = 5V all modes				

BLOCK DIAGRAM

PIN OUT

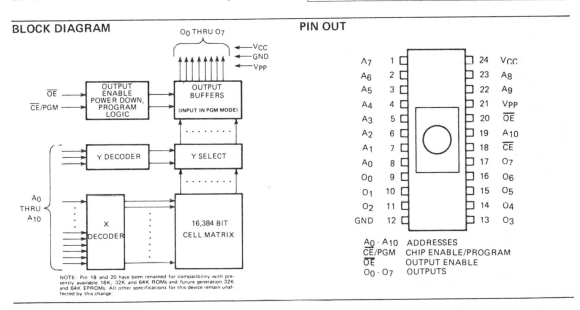

NOTE: Pin 18 and 20 have been renamed for compatibility with presently available 16K, 32K and 64K ROMs and future generation 32K and 64K EPROMs. All other specifications for this device remain unaffected by this change.

A₀ - A10   ADDRESSES
$\overline{CE}$/PGM   CHIP ENABLE/PROGRAM
$\overline{OE}$ OUTPUT ENABLE
O₀ - O7   OUTPUTS

Figure 4.5(a)–(f)   MK2716 EPROM data (Courtesy Mostek)

ABSOLUTE MAXIMUM RATINGS*

Voltage on any pin relative to VSS .... -0.3V to +6V
(Except VPP)
Voltage on VPP supply pin relative to VSS .... -0.3V to +28V
Operating Temperature TA (Ambient) .... $0°C \leq TA \leq 70°C$
Storage Temperature (Ambient) .... $-55°C \leq TA \leq +125°C$
Power Dissipation .... 1 Watt
Short Circuit Output Current 50mA

*Stresses greater than those listed under "Absolute Maximum Ratings" may cause permanent damage to the device. This is a stress rating only and functional operation of the device at these or any other conditions above those indicated in the operating sections of this specification is not implied. Exposure to absolute maximum rating conditions for extended periods may affect device reliability.

READ OPERATION
RECOMMENDED D.C. OPERATING CONDITIONS AND CHARACTERISTICS[1,2,4,8]
$(0°C \leq TA \leq 70°C)$ (VCC = +5V $\pm$5%, VPP = VCC $\pm$ 0.6V)[3]

SYMBOL	PARAMETER	MIN	TYP	MAX	UNITS	NOTES
VIH	Input High Voltage	2.0		Vcc+1	Volts	
VIL	Input Low Voltage	-0.1		0.8	Volts	
ICC1	VCC Standby Power Supply Current ($\overline{OE}$ = VIL; $\overline{CE}$ = VIH)		10	25	mA	2
ICC2	VCC Active Power Supply Current ($\overline{OE}$ = $\overline{CE}$ = VIL)		57	100	mA	2
IPP1	VPP Current (VPP = 5.85V)			5	mA	2,3
VOH	Output High Voltage (IOH = -100 μA)	2.4			Volts	
VOL	Output Low Voltage (IOL = 2.1mA)			.45	Volts	
IIL	Input Leakage Current (VIN = 5.25V)			10	μA	
IOL	Output Leakage Current (VOUT = 5.25V)			10	μA	

A.C. CHARACTERISTICS[1,2,5]
$(0°C \leq TA \leq 70°C)$ (VCC = +5V $\pm$ 5%, VPP = 5V $\pm$ 0.6V)[3]

SYMBOL	PARAMETER	-6		-7		-8		UNITS	NOTES
		MIN	MAX	MIN	MAX	MIN	MAX		
tACC	Address to Output Delay ($\overline{CE}$ = $\overline{OE}$ = VIL)		350		390		450	ns	
tCE	$\overline{CE}$ to Output Delay ($\overline{OE}$ = VIL)		350		390		450	ns	6
tOE	Output Enable to Output Delay ($\overline{CE}$ = VIL)		120		120		120	ns	10
tDF	Chip Deselect to Output Float ($\overline{CE}$ = VIL)	0	100	0	100	0	100	ns	9
tOH	Address to Output Hold ($\overline{CE}$ = $\overline{OE}$ = VIL)	0		0		0		ns	

Figure 4.5(b)

CAPACITANCE
(TA = 25°C)[8]

SYMBOL	PARAMETER	TYP	MAX	UNITS	NOTES
C_{IN}	Input Capacitance	4	6	pF	7
C_{OUT}	Output Capacitance	8	12	pF	7

NOTES:
1. VCC must be applied on or before VPP and removed after or at the same time as VPP.
2. VPP and VCC may be connected together except during programming, in which case the supply current is the sum of I_{CC} and I_{PP1}.
3. The tolerance on VPP is to allow use of a driver circuit to switch VPP from VCC to +25V in the READ and PROGRAM mode respectively.
4. All voltages with respect to VSS.
5. Load conditions = lTTL load and 100pF., tr = tf = 20ns, reference levels are 1V or 2V for inputs and .8V and 2V for outputs.
6. tOE is referenced to $\overline{CE}$ or the addresses, whichever occurs last.
7. Effective Capacitance calculated from the equation $C = \frac{\Delta Q}{\Delta V}$ where $\Delta V = 3V$
8. Typical numbers are for TA = 25°C and VCC = 5.0V
9. tDF is applicable to both $\overline{CE}$ and $\overline{OE}$, whichever occurs first.
10. $\overline{OE}$ may follow up to tACC - tOE after the falling edge of $\overline{CE}$ without effecting tACC

TIMING DIAGRAMS
READ CYCLE ($\overline{CE} = V_{IL}$)

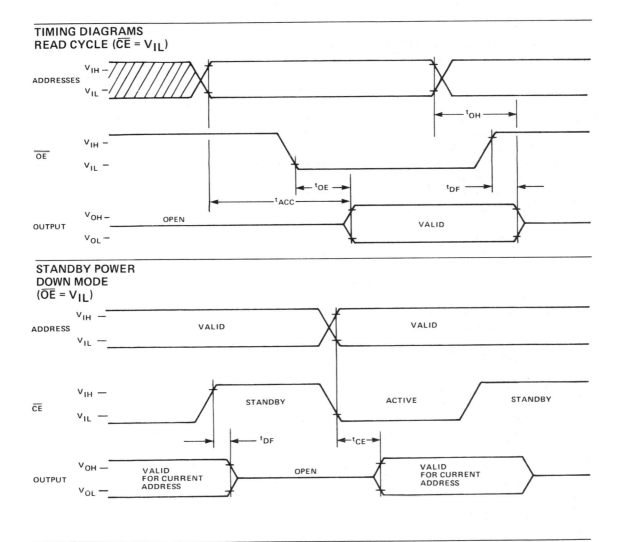

STANDBY POWER
DOWN MODE
($\overline{OE} = V_{IL}$)

Figure 4.5(c)

PROGRAM OPERATION[8]
D.C. ELECTRICAL CHARACTERISTICS AND OPERATING CONDITIONS[1,2]
(TA = 25°C ± 5°C) (VCC = 5V ±5%, VPP = 25V ±1V)

SYMBOL	PARAMETER	MIN	MAX	UNITS	NOTES
IIL	Input Leakage Current		10	μA	3
VIL	Input Low Level	-0.1	0.8	Volts	
VIH	Input High Level	2.0	VCC +1	Volts	
ICC	VCC Power Supply Current		100	mA	
IPP1	VPP Supply Current		5	mA	4
IPP2	VPP Supply Current during Programming Pulse		30	mA	5

A.C. CHARACTERISTICS AND OPERATING CONDITIONS[1,2,6,7]
(TA = 25°C ± 5°C) (VCC = 5V ± 5%, VPP = 25V ± 1V)

SYMBOL	PARAMETER	MIN	TYP	MAX	UNITS	NOTES
tAS	Address Setup Time	2			μs	
tOES	$\overline{OE}$ Setup Time	2			μs	
tDS	Data Setup Time	2			μs	
tAH	Address Hold Time	2			μs	
tOEH	$\overline{OE}$ Hold Time	2			μs	
tDH	Data Hold Time	2			μs	
tDF	Output Enable to Output Float	0		120	ns	4
tOE	Output Enable to Output Delay			120	ns	4
tPW	Program Pulse Width	45	50	55	ms	
tPRT	Program Pulse Rise Time	5			ns	
tPFT	Program Pulse Fall Time	5			ns	

NOTES:
1. VCC must be applied at the same time or before VPP and removed after or at the same time as VPP. To prevent damage to the device it must not be inserted into a board with VPP at 25V.
2. Care must be taken to prevent overshoot of the VPP supply when switching to +25V.
3. 0.45V ≤ VIN ≤ 5.25V
4. $\overline{CE}$/PGM = VIL
5. $\overline{CE}$/PGM = VIH
6. tT = 20nsec
7. 1V or 2V for inputs and .8V or 2V for outputs are used as timing reference levels.
8. Although speed selections are made for READ operation all programming specifications are the same for all dash numbers.

Figure 4.5(d)

TIMING DIAGRAM
(Program Mode)

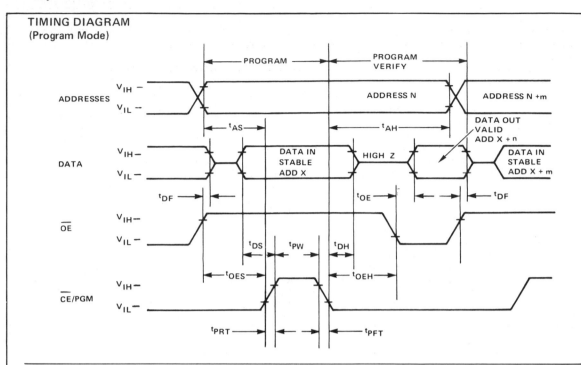

Programming can be done with a single TTL level pulse. and may be done on any individual location either sequentially or at random. The three-state output controlled by the OE input allows OR-tie capability for construction of large arrays. A single power supply requirement of +5 volts makes the MK2716 ideally suited for use with MOSTEK's new 5 volt only microprocessors such as the MK3880 (Z-80). The MK2716 is packaged in the industry standard 24 pin dual-in line package with a transparent hermetically sealed lid. This allows the user to expose the chip to ultraviolet light to erase the data pattern. A new pattern may then be written into the device by following the program procedures outlined in this data sheet.

The MK2716 is specifically designed to fit those applications where fast turnaround time and pattern experimentation are required. Since data may be altered in the device (erase and reprogram) it allows for early debugging of the system program. Since single location programming is available, the MK2716 can have its data content increased (assuming all 2048 bytes were not programmed) at any time for easy updating of system capabilities in the field. Once the data/program is fixed and the intention is to produce large numbers of systems, MOSTEK also supplies a pin compatible mask programmable ROM, the MK34000. To transfer the program data to ROM, the user need only send the PROM along with device information to MOSTEK, from which the ROM with the desired pattern can be generated. This means a reduction in the possibility of error when converting data to other forms (cards, tape, etc.) for this purpose. However, data may still be input by any of these traditional means such as paper tape, card deck, etc.

READ OPERATION

The MK2716 has five basic modes of operation. Under normal operating conditions (non-programming) there are two modes including READ and STANDBY. A READ operation is accomplished by maintaining pin 18 ($\overline{CE}$) at VIL and pin 21 (VPP) at +5 volts. If $\overline{OE}$ (pin 20) is held active low after addresses (A0 - A10) have stabilized then valid output data will appear on the output pins at access time tACC (address access). In this mode, access time my be referenced to $\overline{OE}$ (tOE) depending on when $\overline{OE}$ occurs (see timing diagrams).

POWER DOWN operation is accomplished by taking pin 18 ($\overline{CE}$) to a TTL high level (VIH). The power is reduced by 75% from 525mW maximum to 132mW. In power down VPP must be at +5 volts and the outputs will be open-circuit regardless of the condition of $\overline{OE}$. Access time from a high to low transition of $\overline{CE}$ (tCE) is the same as from addresses (tACC). (See STANDBY Timing Diagram).

PROGRAMMING INSTRUCTIONS

The MK2716 as shipped from MOSTEK will be completely erased. In this initial state and after any subsequent erasure, all bits will be at a '1' level (output high). Information is introduced by selectively programming '0's into the proper bit locations. Once a '0' has been programmed into the chip it may be changed only by erasing the entire chip with UV light.

Word address selection is done by the same decode circuitry used in the READ mode. The MK2716 is put into the PROGRAM mode by maintaining VPP at +25V,

Figure 4.5(e)

and $\overline{OE}$ at VIH. In this mode the output pins serve as inputs (8 bits in parallel) for the required program data. Logic levels for other inputs and the VCC supply voltage are the same as in the READ mode.

The program a "byte" (8 bits) of data, a TTL active high level pulse is applied to the $\overline{CE}$/PGM pin once addresses and data are stabilized on the inputs. Each location must have a pulse applied with only one pulse per location required. Any individual location, a sequence of locations or locations at random may be programmed in this manor. The program pulse has a minimum width of 45msec and a maximum of 55msec, and must not be programmed with a high level D.C. signal applied to the $\overline{CE}$/PGM pin.

PROGRAM INHIBIT is another useful mode of operation when programming multiple parallel addressed MK2716's with different data. It is necessary only to maintain $\overline{OE}$ at VIH, VPP at +25, allow addresses and data to stabilize and pulse the $\overline{CE}$/PGM pin of the device to be programmed. Data may then be changed and the next device pulsed. The devices with $\overline{CE}$/PGM at VIL will not be programmed.

PROGRAM VERIFY allows the MK2716 program data to be verified without having to reduce VPP from +25V to +5V. VPP should only be used in the PROGRAM/PROGRAM INHIBIT and PROGRAM VERIFY modes and must be at +5V in all other modes.

MK2716 ERASING PROCEDURE

The MK2716 may be erased by exposure to high intensity ultraviolet light, illuminating the chip thru the transparent window. This exposure to ultraviolet light induces the flow of a photo current from the floating gate thereby discharging the gate to its initial state. An ultraviolet source of 2537Å yielding a total integrated dosage of 15 Watt-seconds/cm^2 is required. Note that all bits of the MK2716 will be erased. The erasure time is approximately 15 to 20 minutes utilizing a ultra-violet lamp with a 12000μW/cm^2 power rating. The lamp should be used without short wave filters, and the MK2716 to be erased should be placed about one inch away from the lamp tubes. It should be noted that as the distance between the lamp and the chip is doubled, the exposure time required goes up by a factor of 4. The UV content of sunlight is insufficient to provide a practical means of erasing the MK2716. However, it is not recommended that the MK2716 be operated or stored in direct sunlight, as the UV content of sunlight may cause erasure of some bits in a short period of time.

Figure 4.5(f)

pulse width for the programming pulse is 45 ms. Note that the timing diagram (*Figure 4.5(e)*, page 107) is not drawn to scale, since the times t_{AS} and t_{AH} are each only 2 μs. This means that in practice the whole program and program verify cycle take little more time than the width of the program pulse. Therefore, if the program pulse width is typically 50 ms, the whole device can be programmed in about 100 seconds.

If a smaller program is required to be loaded into the EPROM, it is only necessary to program the addresses that actually contain code. For most intelligent EPROM programmers, this would result in a much shorter programming time since they would only program part of the EPROM.

(f) Erasing an EPROM is also an interesting process, and is detailed at the end of the data sheet (*Figure 4.5(f)*). The normal erasure time is noted there to be between 15 and 20 minutes using a typical ultraviolet lamp with a 12 000 mW/cm^2 power rating. This assumes that the EPROM is placed about 1 in (2.5 cm) from the ultraviolet tube.

Most EPROM erasing systems allow a number of EPROMs to be erased simultaneously.

4.4 EEPROM (EAROM) – Electrically Erasable/Alterable Programmable Read Only Memory

The idea of having a Read Only Memory that could retain its data even though the power was removed but could be altered without erasing the complete device has been discussed for a long time. However, it has taken many years of development to achieve such a device which would be compatible with other memories available in microprocessor based systems.

The devices use floating-gate technology, and the process known as **electron tunnelling** to store data in the semiconductor, in much the same way as an **EPROM**. However, many technical problems needed to be resolved including the very small line widths and oxide thicknesses required to produce an efficient storage process. Early devices (EAROMs) used a metal–nitride–

oxide–semiconductor structure. They stored electrons by trapping them within the nitride and oxide dielectrics. However, they had significant disadvantages which included the data disturbance during read operations, the loss of data over time, the requirement for multiple power supplies, one of which was negative, and signal swings beyond normal TTL levels. Further complicating their use was the fact that address and data had to be stable for the entire write cycle which often lasted up to 40 ms. This required extra hardware to interface them to normal microprocessor signals.

Recent developments in the technology have resulted in a great improvement in the techniques employed, such that is now possible to have a pin-compatible EEPROM with a normal static RAM. This means that the parts can be interchanged, but with the difference that the EEPROM will retain its data even when the power is removed.

The operation of the device is as follows. It depends upon an electron-tunnelling mechanism between polysilicon layers within the device. These are shown with their electrical connections in *Figure 4.6*. The diagram shows that the EEPROM contains three polysilicon layers connected to a **select** and **sense** transistor for each bit of memory.

To program a cell electrically, electrons must tunnel on to the floating gate from the first polysilicon layer. This is accomplished by applying a

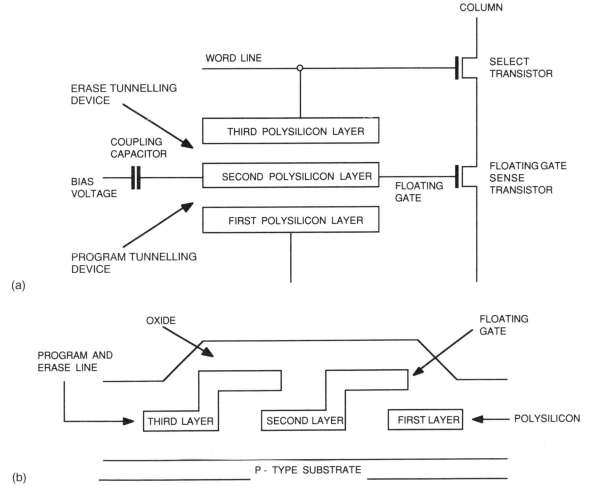

Figure 4.6 EEPROM operation **(a)** Circuit operation **(b)** Physical implementation

| 16K | Commercial
Industrial | X2816B
X2816BI | 2048 x 8 Bit |

Electrically Erasable PROM

FEATURES

- 250 ns Access Time
- High Performance Advanced NMOS Technology
- Fast Write Cycle Times
  - —16-Byte Page Write Operation
  - —Byte or Page Write Cycle: 5 ms Typical
  - —Complete Memory Rewrite: 640 ms Typical
  - —Effective Byte Write Cycle Time of 300 μs Typical
- $\overline{\text{DATA}}$ Polling
  - —Allows User to Minimize Write Cycle Time
- Simple Byte and Page Write
  - —Single TTL Level $\overline{\text{WE}}$ Signal
  - —Internally Latched Address and Data
  - —Automatic Write Timing
- JEDEC Approved Byte-Wide Pinout

DESCRIPTION

The Xicor X2816B is a 2K x 8 E^{2}PROM, fabricated with an advanced, high performance N-channel floating gate MOS technology. Like all Xicor programmable nonvolatile memories it is a 5V only device. The X2816B features the JEDEC approved pinout for byte-wide memories, compatible with industry standard RAMs, ROMs and EPROMs.

The X2816B supports a 16-byte page write operation, typically providing a 300 μs/byte write cycle, enabling the entire memory to be written in less than 640 ms. The X2816B also features $\overline{\text{DATA}}$ Polling, a system software support scheme used to indicate the early completion of a write cycle.

Xicor E^{2}PROMs are designed and tested for applications requiring extended endurance. Data retention is specified to be greater than 100 years.

PIN CONFIGURATIONS

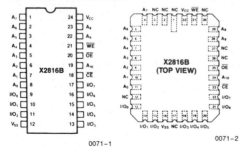

PIN NAMES

A_0–A_{10}	Address Inputs
I/O_0–I/O_7	Data Input/Output
$\overline{\text{WE}}$	Write Enable
$\overline{\text{CE}}$	Chip Enable
$\overline{\text{OE}}$	Output Enable
V_{CC}	+5V
V_{SS}	Ground
NC	No Connect

FUNCTIONAL DIAGRAM

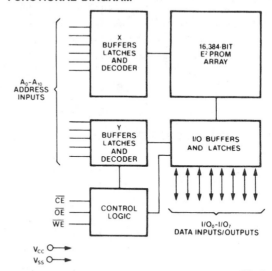

Figure 4.7(a)–(e) X2816 data sheets (Courtesy Xicor)

X2816B, X2816BI

ABSOLUTE MAXIMUM RATINGS*

Temperature Under Bias
X2816B .... $-10°C$ to $+85°C$
X2816BI .... $-65°C$ to $+135°C$
Storage Temperature .... $-65°C$ to $+150°C$
Voltage on any Pin with
Respect to Ground .... $-1.0V$ to $+7V$
D.C. Output Current .... 5 mA
Lead Temperature
(Soldering, 10 Seconds) .... 300°C

*COMMENT

Stresses above those listed under "Absolute Maximum Ratings" may cause permanent damage to the device. This is a stress rating only and the functional operation of the device at these or any other conditions above those indicated in the operational sections of this specification is not implied. Exposure to absolute maximum rating conditions for extended periods may affect device reliability.

D.C. OPERATING CHARACTERISTICS

X2816B $T_A = 0°C$ to $+70°C$, $V_{CC} = +5V \pm 10\%$, unless otherwise specified.
X2816BI $T_A = -40°C$ to $+85°C$, $V_{CC} = +5V \pm 10\%$, unless otherwise specified.

Symbol	Parameter	X2816B/X2816B-25 Limits			X2816BI/X2816BI-25 Limits			Units	Test Conditions
		Min.	Typ.[1]	Max.	Min.	Typ.[1]	Max.		
I_{CC}	V_{CC} Current (Active)		80	120		80	140	mA	$\overline{CE} = \overline{OE} = V_{IL}$ All I/O's = Open Other Inputs = V_{CC}
I_{SB}	V_{CC} Current (Standby)		45	60		45	70	mA	$\overline{CE} = V_{IH}$, $\overline{OE} = V_{IL}$ All I/O's = Open Other Inputs = V_{CC}
I_{LI}	Input Leakage Current			10			10	μA	$V_{IN} = GND$ to V_{CC}
I_{LO}	Output Leakage Current			10			10	μA	$V_{OUT} = GND$ to V_{CC}, $\overline{CE} = V_{IH}$
V_{IL}	Input Low Voltage	-1.0		0.8	-1.0		0.8	V	
V_{IH}	Input High Voltage	2.0		$V_{CC} + 1.0$	2.0		$V_{CC} + 1.0$	V	
V_{OL}	Output Low Voltage			0.4			0.4	V	$I_{OL} = 2.1$ mA
V_{OH}	Output High Voltage	2.4			2.4			V	$I_{OH} = -400 \mu A$

TYPICAL POWER-UP TIMING

Symbol	Parameter	Typ.[1]	Units
t_{PUR}[2]	Power-Up to Read Operation	1	ms
t_{PUW}[2]	Power-Up to Write Operation	5	ms

CAPACITANCE $T_A = 25°C$, f $-$ 1.0 MHz, $V_{CC} = 5V$

Symbol	Test	Max.	Units	Conditions
$C_{I/O}$[2]	Input/Output Capacitance	10	pF	$V_{I/O} = 0V$
C_{IN}[2]	Input Capacitance	6	pF	$V_{IN} = 0V$

A.C. CONDITIONS OF TEST

Input Pulse Levels	0V to 3.0V
Input Rise and Fall Times	10 ns
Input and Output Timing Levels	1.5V
Output Load	1 TTL Gate and $C_L = 100$ pF

MODE SELECTION

$\overline{CE}$	$\overline{OE}$	$\overline{WE}$	Mode	I/O	Power
L	L	H	Read	D_{OUT}	Active
L	H	L	Write	D_{IN}	Active
H	X	X	Standby and Write Inhibit	High Z	Standby
X	L	X	Write Inhibit	—	—
X	X	H	Write Inhibit	—	—

Notes: (1) Typical values are for $T_A = 25°C$ and nominal supply voltage.
(2) This parameter is periodically sampled and not 100% tested.

Figure 4.7(b)

X2816B, X2816BI

A.C. CHARACTERISTICS

X2816B T_A = 0°C to +70°C, V_{CC} = +5V ±10%, unless otherwise specified.
X2816BI T_A = −40°C to +85°C, V_{CC} = +5V ±10%, unless otherwise specified.

Read Cycle Limits

Symbol	Parameter	X2816B-25 X2816BI-25		X2816B X2816BI		Units
		Min.	Max.	Min.	Max.	
t_{RC}	Read Cycle Time	250		300		ns
t_{CE}	Chip Enable Access Time		250		300	ns
t_{AA}	Address Access Time		250		300	ns
t_{OE}	Output Enable Access Time		100		100	ns
t_{LZ}	Chip Enable to Output in Low Z	10		10		ns
t_{HZ}[3]	Chip Disable to Output in High Z	10	60	10	80	ns
t_{OLZ}	Output Enable to Output in Low Z	10		10		ns
t_{OHZ}[3]	Output Disable to Output in High Z	10	60	10	80	ns
t_{OH}	Output Hold from Address Change	10		10		ns

Read Cycle

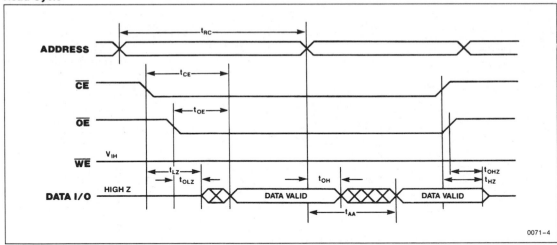

0071-4

Note: (3) t_{HZ} and t_{OHZ} are measured from the point when $\overline{CE}$ or $\overline{OE}$ return high (whichever occurs first) to the time when the outputs are no longer driven.

Figure 4.7(c)

X2816B, X2816BI

Write Cycle Limits

Symbol	Parameter	Min.	Typ.[4]	Max.	Units
t_{WC}	Write Cycle Time		5	10	ms
t_{AS}	Address Setup Time	10			ns
t_{AH}	Address Hold Time	150			ns
t_{CS}	Write Setup Time	0			ns
t_{CH}	Write Hold Time	0			ns
t_{CW}	$\overline{CE}$ Pulse Width	150			ns
t_{OES}	$\overline{OE}$ High Setup Time	10			ns
t_{OEH}	$\overline{OE}$ High Hold Time	10			ns
t_{WP}	$\overline{WE}$ Pulse Width	150			ns
t_{WPH}	$\overline{WE}$ High Recovery	50			ns
t_{DV}	Data Valid			300	ns
t_{DS}	Data Setup	100			ns
t_{DH}	Data Hold	15			ns
t_{DW}	Delay to Next Write	500			μs
t_{BLC}	Byte Load Cycle	3		20	μs

$\overline{WE}$ Controlled Write Cycle

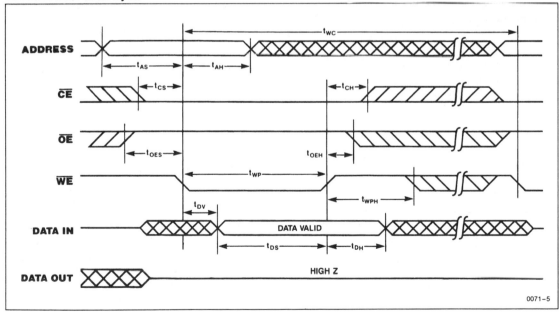

Note: (4) Typical values are for $T_A = 25°C$ and nominal supply voltage.

Figure 4.7(d)

X2816B, X2816BI

PIN DESCRIPTIONS

Addresses (A$_0$–A$_{10}$)
The Address inputs select an 8-bit memory location during a read or write operation.

Chip Enable ($\overline{CE}$)
The Chip Enable input must be LOW to enable all read/write operations. When $\overline{CE}$ is HIGH, power consumption is reduced.

Output Enable ($\overline{OE}$)
The Output Enable input controls the data output buffers and is used to initiate read operations.

Data In/Data Out (I/O$_0$–I/O$_7$)
Data is written to or read from the X2816B through the I/O pins.

Write Enable ($\overline{WE}$)
The Write Enable input controls the writing of data to the X2816B.

DEVICE OPERATION

Read
Read operations are initiated by both $\overline{OE}$ and $\overline{CE}$ LOW. The read operation is terminated by either $\overline{CE}$ or $\overline{OE}$ returning HIGH. This 2-line control architecture eliminates bus contention in a system environment. The data bus will be in a high impedance state when either $\overline{OE}$ or $\overline{CE}$ is HIGH.

Write
Write operations are initiated when both $\overline{CE}$ and $\overline{WE}$ are LOW *and* $\overline{OE}$ is HIGH. The X2816B supports both a $\overline{CE}$ and $\overline{WE}$ controlled write cycle. That is, the address is latched by the falling edge of either $\overline{CE}$ or $\overline{WE}$, whichever occurs last. Similarly, the data is latched internally by the rising edge of either $\overline{CE}$ or $\overline{WE}$, whichever occurs first. A byte write operation, once initiated, will automatically continue to completion, typically within 5 ms.

Page Write Operation
The page write feature of the X2816B allows the entire memory to be typically written in 640 ms. Page write allows two to sixteen bytes of data to be consecutively written to the X2816B prior to the commencement of the internal programming cycle. Although the host system may read data from any location in the system to transfer to the X2816B, the destination page address of the X2816B should be the same on each subsequent strobe of the $\overline{WE}$ and $\overline{CE}$ inputs. That is, A$_4$

through A$_{10}$ must be the same for each transfer of data to the X2816B during a page write cycle.

The page write mode can be entered during any write operation. Following the initial byte write cycle, the host can write an additional one to fifteen bytes in the same manner as the first byte was written. Each successive byte load cycle, started by the $\overline{WE}$ HIGH to LOW transition, must begin within 20 μs of the falling edge of the preceding $\overline{WE}$. If a subsequent $\overline{WE}$ HIGH to LOW transition is not detected within 20 μs, the internal automatic programming cycle will commence. There is no page write window limitation. The page write window is infinitely wide, so long as the host continues to access the device within the byte load cycle time of 20 μs.

$\overline{DATA}$ Polling
The X2816B features $\overline{DATA}$ Polling as a method to indicate to the host system that the byte write or page write cycle has completed. $\overline{DATA}$ Polling allows a simple bit test operation to determine the status of the X2816B, eliminating additional interrupt inputs or external hardware. During the internal programming cycle, any attempt to read the last byte written will produce the complement of that data on I/O$_7$ (i.e., write data = 0xxx xxxx, read data = 1xxx xxxx). Once the programming cycle is complete, I/O$_7$ will reflect true data.

WRITE PROTECTION

There are three features that protect the nonvolatile data from inadvertent writes.

- Noise Protection—A $\overline{WE}$ pulse of less than 20 ns will not initiate a write cycle.

- V$_{CC}$ Sense—All functions are inhibited when V$_{CC}$ is $\leq$3V, typically.

- Write Inhibit—Holding either $\overline{OE}$ LOW, $\overline{WE}$ HIGH or $\overline{CE}$ HIGH during power-on and power-off, will inhibit inadvertent writes.

ENDURANCE

Xicor E^{2}PROMs are designed and tested for applications requiring extended endurance. The process average for endurance of Xicor E^{2}PROMs is approximately $\frac{1}{2}$ million cycles, as documented in RR504, the *Xicor Reliability Report on Endurance*. Included in that report is a method for determining the expected endurance of the device based upon the specific application environment. RR504 and additional reliability reports are available from Xicor.

Figure 4.7(e)

bias voltage to the coupling capacitor to pull the floating gate high capacitively and develop a voltage across the program tunnelling device. This voltage is developed on the chip from the normal 5 V supply and typically is around 20 V. When the voltage is applied electrons tunnel from the first polysilicon layer through the oxide to the second layer and remain there when the program voltage is removed. The floating gate sense transistor is then turned off by the negative voltage produced by the presence of electrons on its floating gate and thus a '0' is produced at the EEPROM output.

To erase the cell electrically electrons must tunnel off the floating gate and this is done again by capacitively coupling the second polysilicon layer to a low voltage while applying a high voltage via the word line to the third polysilicon layer. Electrons then tunnel off the floating gate and when read, the sense transistor conducts producing a '1' at the device output.

The data write operation takes between 5 and 10 ms. This is very slow in comparison with a normal static RAM and about the same time as an EPROM would take to program each byte. The difference is that the EEPROM does not have to be removed from a normal TTL circuit. They are arranged so that the data only needs to be held static for a very short period, like a static RAM, and the device then latches this data while it performs the internal write operation. During the write operation the data may be checked by reading it at any time, and it would be normal for a program to keep reading the data until it matched what was written.

EEPROMs have many applications, and they are likely to become more popular as these applications are developed, and write times are reduced even further. However, small devices can be used for replacing switches and this is already apparent in personal computers and other devices. Wherever a battery backed CMOS RAM chip is used an EEPROM could be a useful substitute. Early devices had only a few bits of storage capability, such as 512 bytes, but modern devices can have up to 256 000. This means that a 32K × 8 EEPROM is a practical proposition. Such devices could have many applications since they could hold a significant amount of software. For example, the prices in point-of-sale terminals or petrol pumps could be retained in EEPROMs and these could be remotely updated with price changes whenever required. There is also another interesting possibility in which it could be possible, for example, to have programs that learn as they are used. For example a video game could become harder by re-programming itself as players got higher scores.

X2816 2K × 8 EEPROM

The following manufacturer's data gives you a good idea of how similar these devices are to normal static RAM. The device chosen is an X2816, which is a 2K × 8 EEPROM.

Main Features

(a) It is pin compatible with the standard byte-wide devices such as the 2716 and the MK4118. This means that it can be interchanged very easily with either device.
(b) Its normal access time for reading is 250 ns and this makes it one of the faster types of Read Only Memory.
(c) The device is fully TTL compatible and requires only a +5 V supply. This contrasts sharply with other earlier devices which also required a +21 V supply.
(d) Data retention is specified as being greater than 100 years.
(e) The data can be checked during the writing operation if the address that has just been written to is read. Until the device is fully programmed, the complement of the data is returned, but when the programming cycle is complete the true data is returned. This allows a simple check without the use of interrupts or any other devices so that no additional pins are required, but only a simple software polling algorithm.

Summary

This chapter has described the main features of typical memory devices to be found in microprocessor based systems. It has provided examples of manufacturer's data sheets which illustrate the level of detail that must be supplied for each component.

- PROMs, EPROMs, EEPROMs and RAMs exist in byte-wide configurations, using the same address and data lines to provide compatibility between devices.
- EPROMs and PROMs tend to have slower access times than RAMs.

- Dynamic RAM devices require a multiplexed address bus to reduce the pin count and hence the size of the chip.

- Dynamic RAMs tend to require multiple power supply lines whereas other devices generally use only a +5 V supply.

Questions

4.1 What is the normal power dissipation of the MK4118 memory chip in its active mode?

4.2 What is the absolute maximum logic '1' input voltage that may be applied to the MK4118 chip?

4.3 In the **write** cycle of a MK4118–1, for how long must the data on the data bus be valid before the $\overline{\text{WE}}$ signal rises to its logic '1' state, to ensure the data is accurately written into the memory?

4.4 Is the MK4116 a TTL-compatible device?

4.5 A 2716 EPROM has been programmed incorrectly in only one byte of its program. The data that was written was 55 hex, but it should have been 54 hex. Describe how this data may be changed.

4.6 A 25 V power supply is being designed to be used in an EPROM programmer capable of programming one device at a time. What would be a sensible current capability for this power supply?

4.7 How is an MK36000 switched into its standby mode?

4.8 How long would it take to completely refresh a MK4116 dynamic memory?

4.9 Briefly explain the difference between the output enable $\overline{\text{OE}}$ and the chip select $\overline{\text{CS}}$ pins on a MK4118 static RAM.

FIVE

Interrupts

When you have finished this chapter, you should be able to:

1. Deduce why interrupts are necessary especially in the handling of data transfers between peripheral and computer.

2. Explain that an interrupt may cause the main program to **call** an 'interrupt service routine' (ISR).

3. Infer that in returning from the ISR, the main program should continue as though it had never been interrupted.

4. Explain the use of the **stack** in saving and restoring MPU registers when servicing an interrupt.

5. Explain the mechanism of the microprocessor response upon receipt of an interrupt.

6. Distinguish between maskable and non-maskable interrupts.

7. Discuss the function and operation of a counter timer chip.

5.1 SYNCHRONISING DATA INPUT AND OUTPUT

One of the most widely known features of computer systems is that they operate at very high speed. Typically a microprocessor operates each instruction in about 2 µs, giving a typical execution speed of 500 000 instructions per second. Unfortunately there are not many peripheral devices that can operate at such a speed and the problem of synchronising data input and output has to be addressed.

Typical computer peripherals include keyboards and printers and these devices demonstrate the problem. A typical keyboard, even with a relatively fast typist operating it, cannot input data at more than about five characters per second. When a printer is attached to a computer

the data output is ultimately determined by a speed of the print head. On a typical matrix printer this may be about 100 characters per second, or slightly higher on some of the more modern machines. Even this, however, is very different to the speed of operation of the microprocessor.

Since there is such a variety in speed, it is apparent that one microprocessor should be able to service a number of peripherals, giving each one of them very rapid response, and making them appear to have exclusive use of the processor. Even when all the data input and output transfers are completed, the processor should still have time to perform the calculations and other operations required by the program that is being operated. This is the basis of a multiuser computer system, and also the basis of most systems that include more than one input or output device.

Computers can handle input and output in a number of different ways, the simplest of which is simply to wait for the peripherals to supply data. Unfortunately this method has a serious disadvantage, since the processor time is nearly all wasted waiting for the very slow peripheral, and the computer is therefore effectively operating at the peripheral speed.

Two much more satisfactory methods of synchronisation are used widely in computer systems and these are known as **polling**, and the use of **interrupts**.

5.2 POLLING

Polling is a technique used in many multiuser computer systems, in which the CPU regularly

117

checks each of the peripheral ports to see whether data is required, or whether input data is available. It normally does this as a separate subroutine, which is executed regularly no matter what other program the machine is running. In this way, for example, each peripheral may be checked up to ten times per second for data or at a speed that relates to the maximum data rate of the peripheral.

Polling takes place under the control of a 'polling algorithm', a typical example of which is shown in *Figure 5.1*. The polling algorithm warrants a little further investigation. The flow chart appears to be a number of tests, which branch off to the peripheral service routine if the test proves that a data transfer is required. If no service is required then each test simply passes on to the next one. This means that peripherals are always serviced in the same order, depending upon their order in the polling routine. It also means that each peripheral is tested in turn, whether or not data is available and this affects the time taken for the polling routine.

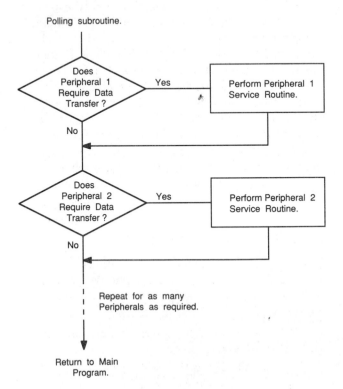

Figure 5.1 Typical polling algorithm

For example, the test to determine whether data is required to be transferred, could be a number of microprocessor operations. If the peripheral is connected via a serial input/output device, the test would first need to read in the status of the input register, to determine whether a character had been received. One of the bits would then be tested. If the test showed that a character had been received the service routine would be executed. This in turn would read in the data from the data register of the peripheral, store it in the memory location and then return to the polling algorithm. Action would only be taken as a result of the character input when the complete polling algorithm was complete. If every peripheral required a data transfer, clearly the polling routine would take much longer than the situation when no peripherals required any data transfer.

Different polling algorithms are possible for different hardware configurations. If a number of ports are connected so that each one has a status flag which is a single bit of a combined register, then by reading the status register only once, it would be possible to test whether data was available in every port connected to it. The test for each peripheral would then simply be a matter of reading the status register and checking for any bits that were set. A bit that was set then would indicate which peripheral required the data transfer.

The main features of polling are as follows:

(a) It is initiated by the CPU, through the software.

(b) Each peripheral needs to be checked sufficiently frequently so that data is not lost.

(c) The peripherals must not be checked so frequently that the CPU has no time to do anything else.

(d) Priorities can be assigned to certain peripherals by checking them more frequently than others.

(e) Because polling is a regular event, it is best suited to peripherals which supply data in a relatively regular fashion.

5.3 INTERRUPTS

An **interrupt**, as the name suggests, is a signal from a peripheral device which is used to interrupt the main computer program, and force the CPU to transfer data between itself and the peripheral. The intention is to perform the data transfer, and then allow the CPU to continue with the program it was running before the interrupt occurred, as though nothing had happened. When this occurs, the CPU is said to leave its main program and perform an 'interrupt service routine'.

Since an interrupt could occur at any point in the operation of the main program, it is vital that the service routine results in the CPU being left in exactly the same state as it was before the interrupt occurred. The only changes that must take place as a result of this service routine are the transfer of data between the peripheral and a memory location or an otherwise unused CPU register.

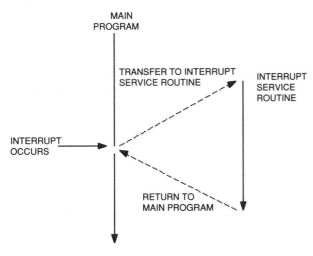

Figure 5.2 Interrupt service routine

There are two main types of interrupt – **maskable** and **non-maskable**. The difference between them is that a maskable interrupt can be prevented by the use of appropriate software instructions, whereas a non-maskable interrupt cannot be prevented by the computer software. As such, non-maskable interrupts tend to be used for high priority signals whereas the normal maskable interrupts are used for the connection of peripheral devices.

The interrupt signal is a hardware logic level generated by the peripheral device which is directly attached to a CPU pin. Its effect is to stop the current program being operated, and this occurs as follows:

(a) At the end of each instruction the processor checks to see whether an interrupt signal has been generated. If it has, the processor performs an operation that is designed to determine the start address of the interrupt service routine. The precise method used by the CPU varies from processor to processor, and between different modes operated by the same CPU. The Z80 modes are discussed in more detail later.

(b) When the interrupt service routine start address is determined, the processor **calls** that address. This has the effect of saving the value of the current program counter on the stack.

(c) The interrupt service routine then saves the contents of the registers currently being used by the CPU and these are also placed on the stack.

(d) The service routine input/output operation then takes place.

(e) The register contents previously stored on the stack are restored.

(f) The end of the service routine contains a return from interrupt instruction which forces the processor to return to the main program, as though it had not been interrupted.

It can be seen from the preceding description that quite a number of operations must take place before the actual input or output of data happens. This time delay may be critical in some instances, but it is relatively fast compared with the alternative method of polling an input to see if data is available. Polling involves a regular checking of the inputs and therefore on average, there will be a significant time delay between data being available and data being input or output. With an interrupt driven system, the peripherals receive immediate attention although there is a small time delay built into the process.

None the less there are some issues related to the speed of operation, which may make interrupts an undesirable option. For example, if the processor is performing a critical timing operation, this would be extended if an interrupt was allowed to occur during its execution. Therefore timing operations are best performed either when interrupts are disabled, or by external circuits such as a CTC (counter timer chip).

In addition, since a processor may have many peripherals, the use of interrupts raises the question of how to handle multiple simultaneous interrupts. The normal method of dealing with these is to assign each one a particular **priority**. This means that certain peripherals have precedence over others, and this has to be arranged with hardware logic circuits. Different CPUs have different methods of achieving this. For example, the 8085 has multiple interrupt pins of differing priorities. The Z80 uses a unique 'daisy chain' system which is discussed later. It is also possible to obtain interrupt encoder chips which will allow logic circuits to prioritise interrupt signals (*Figure 5.3*). Not only must the processor be capable of distinguishing the highest priority in a multiple interrupt situation, but it must also be capable of locating the correct interrupt service routine for each of the peripherals.

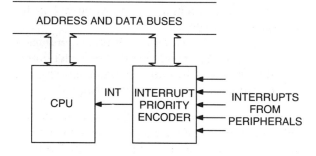

Figure 5.3   Interrupt priority circuit

Interrupt Service Routines

In many ways an interrupt service routine is similar to a normal subroutine:

(a) An interrupt service routine is initiated by an external device which forces a **call** to another routine. A normal subroutine is initiated by a program instruction in much the same way.

(b) The **call** mechanism in each case stores the current program counter contents on the stack so that the routine may return to the same point from which it left the main program.

(c) Both the subroutine and an interrupt service routine must be terminated by a return instruction. In the case of a Z80 interrupt service routine, a special return instruction must be used.

(d) In both cases, data used in the main program may be stored on the system stack.

Therefore it can be seen that an interrupt service routine is essentially the same as a subroutine which is initiated by an external stimulus.

Figure 5.4 shows the detailed content of a typical interrupt service routine.

If maskable interrupts are being employed, the interrupt will only be accepted by the CPU when an enable interrupt instruction has been executed. This has the effect of setting an interrupt flip-flop in the CPU which allows interrupts to be accepted. An interrupt may occur at any point after the execution of this instruction.

The sequence of events is then as follows:

(a) When the interrupt occurs the processor completes the current instruction, then waits to receive the address of the interrupt service routine. There are a number of ways in which this can be achieved, which will be dealt with later.

(b) When the interrupt service routine start address is determined, the processor pushes the current program counter contents onto the stack, then jumps to the beginning of the service routine. This is effectively the same as a call instruction.

(c) At the same time it disables the interrupt flip-flop which effectively prevents further interrupts from occurring until another enable interrupt instruction is executed. If an enable interrupt instruction is omitted from the service routine, then no further interrupts would ever be accepted.

(d) Normally the beginning of the interrupt service routine contains a series of instructions which save the contents of the CPU registers on the stack. These are usually **push** instructions. Clearly, the instructions must be at the beginning of the service routine, and cannot

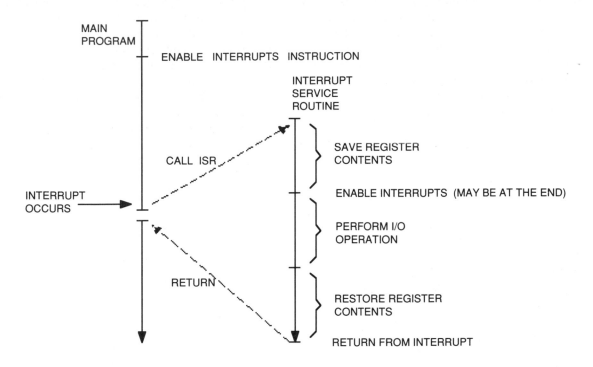

Figure 5.4 Content of an interrupt service routine

be in the main program as with a **call** instruction, since it is not known when the interrupt will occur.

(e) Following the instructions that save the register contents, and the enable interrupt instruction the normal input/output (I/O) operation then takes place, which is the function of the interrupt. This I/O operation **reads** data from the port and places it in memory, or **writes** data to the port from memory. In addition it may perform some operation on the data such as simple code translation or masking certain bits.

(f) When the I/O operation is complete, the data must be restored to the CPU registers using **pop** instructions, so that it may continue with its main program. Occasionally the enable interrupt instruction is placed after the data is restored to the CPU registers. The effect of this is to disable interrupts for the whole duration of the service routine, thus affording it a high priority.

(g) The final instruction in the interrupt service routine must be a return which will restore control to the main program and allow it to continue as though an interrupt had not occurred.

5.4 Z80 INTERRUPTS

Since this book has concentrated on the Z80 microprocessor, it will be quite instructive to examine in detail the types of interrupts found in a Z80-based system. This is not as restricted as it may seem, since the range of interrupts found in a Z80 system is typical of the majority found in all other systems.

Types of Interrupt

Although the concept of using an interrupt is relatively simple, the range of types of interrupt and the variety of names given by different manufacturers tends to complicate the issue.

Interrupts can be divided into a number of types, as *Figure 5.5* shows. The modes shown refer to the Z80 Interrupt system. The first distinction between different types of interrupt is into the types maskable and non-maskable.

> Non-maskable interrupts **cannot** be prevented by software.
>
> Maskable interrupts **can** be prevented by software.

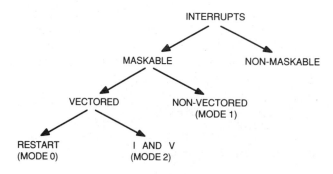

Figure 5.5   Types of interrupt

Non-Maskable Interrupts

Non-maskable interrupts are signals that are sent to the $\overline{\text{NMI}}$ pin of the Z80. Whenever this pin is made to go to a logic 0, the Z80 responds to the interrupt at the end of its current instruction. Its response is very simple. It performs a **restart** to address 0066 hex. A **restart** is just like a **call** instruction but its address is fixed. The contents of the program counter are pushed onto the stack before the program jumps to the new address. When the interrupt service routine is completed for the non-maskable interrupt, control must be passed back to the main program by a 'return from non-maskable interrupt' instruction (RETN).

Non-maskable interrupts tend to be used for special conditions in microprocessor systems. If the microprocessor controls a computer with hard-disk storage, the $\overline{\text{NMI}}$ pin may be connected to a power supply monitor circuit. If the power supply should fall below its normal value, the circuit can cause an interrupt which will then store the current memory contents on disk before the power disappears altogether.

Another common use is in the provision of a **single step** facility. If the processor receives an interrupt after each instruction is read, then it can be made to step through a program one instruction at a time. This, however, requires both a suitable program and some additional hardware to implement it properly.

Processors other than the Z80 normally have a non-maskable interrupt pin but it may have a different name, such as the **trap** input pin of an 8085.

Maskable Interrupts

Maskable interrupts are signals which are sent to the $\overline{\text{INT}}$ pin on the Z80. Before the Z80 will respond to such a signal it must have already executed the enable interrupt (EI) instruction. When the processor is first switched on, the interrupts are disabled and so before they can be used they must be specifically enabled. Whenever the Z80 responds to an interrupt, subsequent interrupts are **disabled** until the processor executes another EI instruction. If a critical part of the program is about to be executed which must not be interrupted, the maskable interrupts can be prevented by including the disable interrupts (DI) instruction in the program.

Interrupt Modes

When the Z80 receives an interrupt signal on the $\overline{\text{INT}}$ pin, its response is determined by the interrupt mode which it has been programmed to expect. There are three possible modes: mode 0, mode 1 and mode 2.

An interrupt mode is set by executing one of the 'set interrupt mode' instructions, IM0, IM1 or IM2. One of these instructions must be executed before a maskable interrupt is received by the Z80 or it will not know how to respond.

The three interrupt modes fall into two categories: **vectored** and **non-vectored**.

Non-Vectored Interrupts

Non-vectored interrupts are those that cause the processor to go to a specific address (as with the non-maskable interrupts). Interrupt mode 1 programs the Z80 to perform a non-vectored interrupt. It responds to the $\overline{\text{INT}}$ signal by performing a **restart (call)** to address 0038 hex. The interrupt service routine must start at this address and it must be terminated with a 'return from interrupt' (RETI) instruction.

Vectored Interrupts

A vectored interrupt allows the processor to select one of a number of start addresses for the interrupt service routine. The Z80 finds the required interrupt service routine in one of two ways.

In mode 0, the peripheral places a single byte on the data bus when the interrupt is acknowledged and using this, the CPU may select one of a number of addresses for the start of the service routine.

In mode 2, the peripheral also places a single byte on the data bus when the interrupt is acknowledged, but it is used in a totally different way to find the start of the required service routine.

Mode 0 Interrupts

Mode 0 interrupts are included in the Z80 instructions set so that it maintains compatibility with the Intel 8080 microprocessor.

Consider the system shown in *Figure 5.6*. Associated with each input port which requires to use mode 0 interrupts, there must be some interrupt logic, which generates the $\overline{\text{INT}}$ signal. There must also be an extra port, called the **restart port** in *Figure 5.6* which is generally wired to a fixed binary number. This number is the binary code for one of the **restart** instructions in the Z80 instruction set. For example:

Restart   0 – C7 hex
Restart   8 – CF hex
Restart 16 – D7 hex etc.

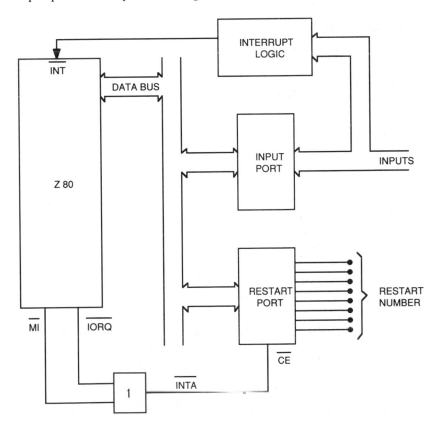

Figure 5.6   Mode 0 interrupts

The system works as follows:

(a) When the inputs to the input port are at a predetermined logic state, the interrupt logic generates a logic 0 on the INT pin.

(b) If the Z80 has been programmed to expect mode 0 interrupts and the interrupts have been **enabled**, it completes its current instruction, then sends out an interrupt acknowledge signal. This consists of both MI and IORQ going low at the same time.

(c) The interrupt acknowledge signal is used as the chip enable input to the **restart** port which then places its **restart number** on the data bus.

(d) The restart number is interpreted by the Z80 as an instruction, which it then executes. Note that this is the only occasion when an instruction comes from anywhere other than a memory chip.

(e) Since the restart number is effectively a 1-byte call, the processor begins the subroutine at the address which corresponds with the number. For example:

> Restart 0 (C7 hex) – Address 0000 hex
> Restart 8 (CF hex) – Address 0008 hex

Note that there are only 8 bytes of memory between each of the addresses which may be called in this way, and this is clearly far too few for a complete service routine. Therefore it is common practice to put a **jump** instruction at the restart address so that the true interrupt service routine starts at a more convenient memory location.

The subroutine may be as long as necessary at the new location, but when it is completed, the last instruction must be a normal **return** instruction (C9 hex). In this respect, mode 0 interrupts differ from mode 2 interrupts which have a special **return from interrupt** instruction to terminate the interrupt service routine.

Mode 0 interrupts are designed to be 8080 compatible and, therefore, will work with relatively unsophisticated hardware. The extra circuits shown in *Figure 5.6* represent a hardware overhead but they are all very simple TTL logic circuits which are relatively inexpensive.

Mode 2 Interrupts

Mode 2 interrupts are designed particularly for the Z80 family of integrated circuits such as the Z80 PIO and the Z80 CTC. They offer a faster and much more flexible method of organising the interrupts in a system, although initially their operation appears more complex.

Consider the situation where a Z80 CPU is connected into a system which contains a Z80 PIO chip and a memory chip. This is shown in *Figure 5.7*.

A number of things must be set up by the main program before mode 2 interrupts can operate. This would normally be carried out by an initialisation program.

Assume that the service routines for ports A and B of the PIO start at addresses 1910 hex and 1A38 hex respectively. (Note that these values are examples only.)

There must also be, somewhere in memory, a 'service routine start address table'. This is shown in the diagram at address 1850 hex and it contains the start addresses of the two service routines that will be used. If more peripheral devices can cause an interrupt, then there will be more service routines, and more entries in the service routine start address table. They generally follow in consecutive addresses, with the low byte of each 'start address' first.

The Z80 CPU contains an interrupt register I. This can only be programmed from the accumulator (see the 8 bit **load** group of instructions) and it must be loaded with the **high byte** of the address of the entry in the service routine start address table relevant to the port. For example, vector V_A is programmed with 50 hex because the start address of the service routine for Port A is found in address 1850 (and 1851).

Note that the only restriction on the location of the service routine start address table is that it must start on an **even** number.

When all of these values have been set up by the system **initialisation** program, the **interrupts** can be **enabled**. Suppose that the input data to port A of the PIO is such that an interrupt is generated. The system works as follows:

(a) The input data to port A causes an interrupt signal INT to be sent to the CPU.

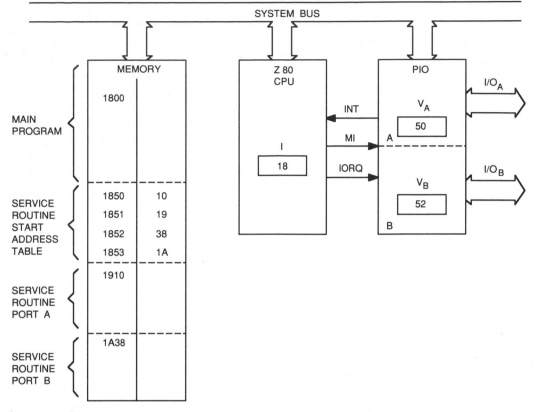

Figure 5.7 Mode 2 interrupts

(b) If interrupts are **enabled**, the CPU completes its current instruction then acknowledges the interrupt by making $\overline{\text{MI}}$ and $\overline{\text{IORQ}}$ low simultaneously.

(c) The PIO places its vector from port A (50 hex) on the data bus and this is read by the CPU.

(d) The CPU combines its I register contents and the vector V to form an address (in this example, 1850 hex).

(e) The CPU reads the numbers in addresses 1850 and 1851 hex and treats them as the start address of the interrupt service routine (in this example, 1910 hex).

(f) The CPU then **calls** the address found in the service routine start address table and hence goes directly to the interrupt service routine.

When the service routine is completed, the last instruction must be a **return from interrupt (RETI)** which has the effect of resetting the PIO interrupt logic as well as performing a normal return operation.

Mode 2 interrupts have a number of advantages over mode 0 interrupts for Z80 based systems.

(a) The interrupt response is faster even though it appears more complex.

(b) Many more interrupts can be accommodated because there is no restriction such as the number of **restart** instructions.

(c) Service routines can start anywhere in memory, not just at fixed addresses.

(d) No extra hardware is required for Z80 family devices.

Daisy Chain System

The **daisy chain** system is the means used by Z80 family chips to achieve a **priority** structure for the **interrupts** which they generate. Its main feature is that all of the logic circuits required are built into the chips and, therefore, no external logic devices are required.

Each device in the Z80 family has two special

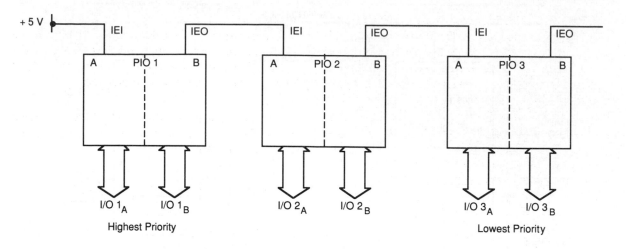

Figure 5.8 Daisy chain priority scheme

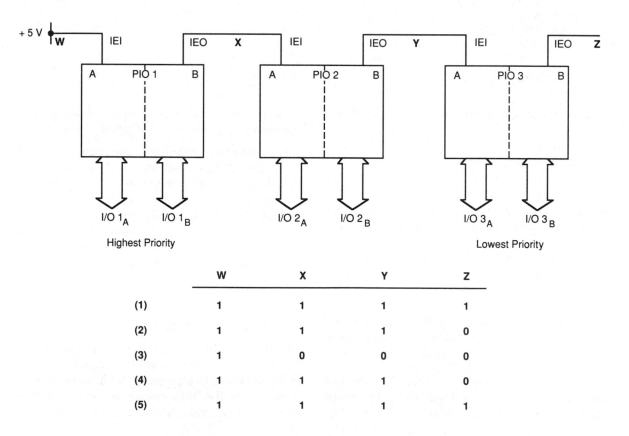

	W	X	Y	Z
(1)	1	1	1	1
(2)	1	1	1	0
(3)	1	0	0	0
(4)	1	1	1	0
(5)	1	1	1	1

Figure 5.9 Daisy chain signals

pins known as **interrupt enable IN** (IEI) and **interrupt enable OUT** (IEO). When the system is designed these pins are wired together like a 'daisy chain'. The first device in the chain has the **highest** priority and devices further along the chain have progressively **lower** priorities. This is illustrated in *Figure 5.8*.

If the IEI pin of a device is at logic 0, its IEO pin is forced to a logic 0 also. When the IEI pin is a logic 1, the IEO pin is also a logic 1, unless the port is interrupting the CPU, when it becomes a logic 0.

Within a PIO chip, port A has a higher priority than port B. The highest priority device has its IEI pin connected to a logic 1 (+5 V).

The system is arranged so that the lower priority devices can have their interrupt service routines interrupted by the higher priority devices. However, low-priority devices cannot interrupt the service routines of the higher priority devices. This assumes that the enable interrupt instruction is executed near the beginning of each interrupt service routine.

The sequence of events that takes place when first PIO 3A generates an interrupt, then PIO 1A generates an interrupt is illustrated in *Figure 5.9*:

- *Stage (1)*
  – illustrates the daisy chain signals before any interrupts occur. All IEI and IEO pins are at logic 1.
- *Stage (2)*
  – illustrates the signals just after PIO 3 has generated an interrupt. Its IEO pin is changed to a logic 0. Note that this is the only PIO with IEI at logic 1 and IEO at logic 0. Because of this it accepts the interrupt acknowledge signal sent from the CPU, and proceeds with its service routine.
- *Stage (3)*
  – illustrates the condition when the higher priority PIO 1 has just interrupted the service routine of PIO 3. It forces its IEO pin to logic 0 and this forces all the lower priority devices IEI and IEO pins to logic 0. PIO 1 then has its IEI pin at logic 1 and IEO pin at logic 0 so that it receives the interrupt acknowledge from the CPU and proceeds with its service routine. The service routine of PIO 3 is temporarily suspended.

- *Stage (4)*
  – illustrates the condition when PIO 1 service routine has been completed. It then allows the PIO 3 service routine to be resumed and eventually this is also completed.
- *Stage (5)*
  – illustrates the condition when PIO 3 service routine is complete and the system is waiting for another interrupt.

Programming for Z80 Mode 2 Interrupts

A typical interrupt may be generated by data changing at a PIO input pin.

Before an interrupt can be accepted by the Z80, a number of events must take place which initialise both the CPU and PIO. These are illustrated in *Figure 5.10*.

The method of initialising the CPU has been dealt with already, but the process of initialising the PIO needs to be examined further.

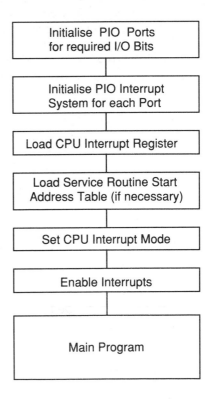

Figure 5.10 Initialisation flow chart

PIO Interrupt Initialisation

PIO initialisation is achieved by sending a number of bytes of data to the **control** port associated with each **data** port. Each port of data has a specific function and must be constructed very carefully, generally in **binary** form.

The first control byte sent to a port usually sets up its operating **mode**, and the second selects the required **input** or **output** bits if mode 3 (bit control) is selected.

For example, the program below sets up port 80H with bits 0, 1 and 2 as inputs and bits 3–7 as outputs, by sending the control bytes to port 82H:

```
        ORG 1800 H
INIT:   LD A,OFFH
        OUT (82H),A   ;  SET MODE 3
        LD A,07H
        OUT (82H),A   ;  BITS 0–2 IN
```

If **interrupts** are required, further control bytes must be sent as follows.

(a) Interrupt vector (*Figure 5.11*) The interrupt vector always end in a 0 in bit 0. Often the interrupt vector is sent as the first control byte to a port even before the **mode** control byte. This vector represents the **low** part of the address where the address of the interrupt service routine will be found.

V7	V6	V5	V4	V3	V2	V1	0

Figure 5.11   Interrupt vector

(b) Interrupt control byte (*Figure 5.12*) Each bit of this byte has a special purpose and its construction must be carefully considered.

Mask follows (MF) – If an interrupt may be generated by some of the PIO inputs changing state but not others, then the interrupt control byte must be followed by an interrupt mask. This mask must be sent to the PIO immediately after the interrupt control byte and the fact that it will be sent is indicated by putting a logic 1 in the mask follows bit. With a logic 0 in this bit all inputs may generate an interrupt.

High or low logic (H/L) – The PIO may be programmed to expect active high or active low signals to produce the interrupt. The active level of these inputs is generally dependent upon the design of the external circuits.

AND or OR Logic (A/O) – When there are a number of inputs thath can cause an interrupt, they may be combined in one of two ways. If **AND** logic is selected, then **all** the inputs must be in the active state before the interrupt is generated. If **OR** logic is selected, any one of the inputs in the active state will generate the interrupt.

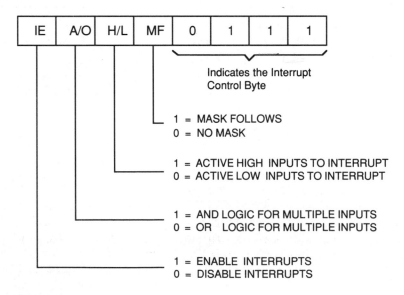

Figure 5.12   Interrupt control byte

Interrupt enable (IE) – Each PIO must have its interrupt logic enabled before it can cause an interrupt. This is in addition to the CPU interrupt enabling with the interrupt enable instruction.

(c) Interrupt mask (*Figure 5.13*) If it is necessary to send an interrupt mask then its byte must follow immediately after the interrupt control byte. The PIO and CPU initialisation process may be clarified by two short examples.

M7	M6	M5	M4	M3	M2	M1	M0

Each bit M0 - M7 may be a logic 0 or logic 1.

M = 1 Ignore   bit for Interrupt purposes
M = 0 Monitor bit for Interrupt purposes

Figure 5.13   Interrupt mask

EXAMPLE 1
Consider a PIO which must be initialised so that bits 2, 3 and 4 are inputs and all other bits are outputs. A high logic signal on bit 3 or 4 may cause an interrupt, and the interrupt vector is 66 hex. What bytes must be sent to the control port?

Solution
(a) Set the P10 to mode 3:

$$1\ 1\ X\ X\ 1\ 1\ 1\ 1 \quad = \quad \text{FF hex}$$

(b) _____
$$0\ 0\ 0\ 1\ 1\ 1\ 0\ 0 \quad = \quad \text{1C hex}$$

(c) Send the interrupt vector:

$$0\ 1\ 1\ 0\ 0\ 1\ 1\ 0 \quad = \quad \text{66 hex}$$

(d) Send the interrupt control byte.
    Enable interrupt, OR logic, active HIGH and MASK follows

$$1\ 0\ 1\ 1\ 0\ 1\ 1\ 1 \quad = \quad \text{B7 hex}$$

(e) Send the interrupt mask. Monitor bits 3 and 4:

$$1\ 1\ 1\ 0\ 0\ 1\ 1\ 1 \quad = \quad \text{E7 hex}$$

These bytes are sent in sequence to the PIO control port.

EXAMPLE 2
A set of eight switches is connected to port 80H in a system, and a set of eight lights is connected to port 81H.

Write a program, including CPU and PIO initialisation which operates as follows.

The main program simply inputs the data from bits 4–7 of the input port and displays the contents on the corresponding lights of the output port. However, when any one of bits 0–3 of the input port is taken to a logic 0, an interrupt is generated which turns the lights of bits 0–3 of the output port on and turns the others off.

The initialisation program starts at address 1800 hex, the interrupt service routine at address 1850 hex and the service routine start address table at address 1840 hex.

The PIO control port for port A is number 82 hex, and for port B it is 83 hex.

Solution
The program begins by initialising the ports of the PIO. Port A requires additional control bytes since it will be used to generate the interrupt signals.

When the PIO is initialised, the CPU and ISR start address tables are set up, and interrupts are enabled.

The main program is an endless loop, which can be interrupted by the generation of the appropriate input signal. When this happens the service routine is executed.

Note that since the accumulator is used in the **main** program to store input data, its contents must be preserved by the service routine. This is achieved by pushing AF onto the **stack** and then popping AF at the end of the service routine.

PROGRAM 1

```
            ORG 1800H
INITA:      LD A,40H                ;   Interrupt vector
            OUT (82H),A             ;   Load interrupt vector
            LD A,0FFH
            OUT (82H),A             ;   MODE 2 Selected Port A
            OUT (82H),              ;   All bits INPUTS
            LD A,97H                ;   10010111
            OUT (82H),A             ;   Load interrupt control byte
            LD A,0F0H               ;   1111 0000 – Mask
            OUT (82H),A             ;   Monitor bits 0–3 only

;
INITB:      LD A,0FFH
            OUT (83H),A             ;   Mode 3 selected port B
            LD A,00
            OUT (83H),A             ;   All bits OUTPUTS

;
CPUINI:     LD A,18                 ;   Interrupt service routine start
            LD I,A                  ;   Address table, high byte in I
            LD HL,1850H             ;   ISR start address
            LD (1840H),HL           ;   Load start address table
            IM 2                    ;   Mode 2 interrupts
            EI                      ;   Enable interrupts
;
MAIN:       IN A,(80H)              ;   Get input data
            AND 0F0H                ;   Mask bits 0–3
            OUT (81H),A             ;   Output to lights
            JP MAIN                 ;   Loop endlessly
```

Note that the service routine starts at address 1850 hex.

```
            ORG 1850H
ISR:        PUSH AF                 ;   Save accumulator
            LD A,0FH
            OUT (81H),A             ;   Turn lights 0–3 ON, others OFF
            LD BC,0000H             ;   Short delay
DEL:        DEC BC
            LD A,B
            OR C
            JP NZ,DEL
            POP AF                  ;   Restore accumulator
            EI                      ;   Enable interrupts again
            RETI                    ;   Return from interrupt
```

5.5 THE Z80 COUNTER TIMER CHIP (CTC)

The Z80 counter timer chip is another dedicated circuit in the Z80 family which can be used to generate interrupts to the Z80 processor under certain circumstances. Its primary functions are twofold: first to perform specific timing operations, and second to perform counting functions. It is generally used in circumstances that would relieve the processor of these mundane tasks. It is a very versatile device, and must be initialised in

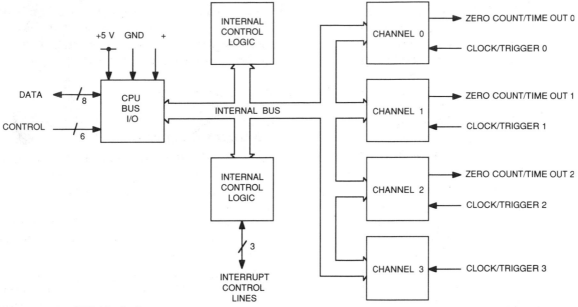

Figure 5.14 CTC block diagram

much the same way as the Z80 PIO device. Although it is relatively complex, the programming is straightforward once its mode of operation is understood.

CTC Architecture

The CTC block diagram is shown in *Figure 5.14*. Its internal data bus interconnects all the channels and connects with the system data bus via a bus interface. Channels are identified with the num-

bers 0–3. Each one may generate a unique interrupt vector when used in an interrupt-driven system, with channel 0 having the highest priority. Each channel has four main parts. The structure of each channel is the same as that shown in *Figure 5.15* and consists of two registers, two counters and some logic.

The registers are the channel control and the time constant registers each having eight bits. The counters are known as the prescaler and the down counter and also have 8 bits.

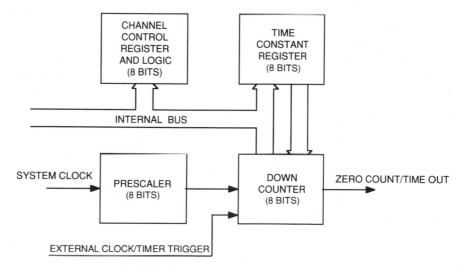

Figure 5.15 Channel block diagram

Each CTC channel may be programmed to operate either as a **counter** of external events via the external clock input, or as a **timer**, using the system crystal clock as the reference input.

The Channel Control Register

Before the CTC channel can operate, its channel control register must be programmed. Each bit of the data byte sent to this register sets up one of its operating conditions. For example, the contents of this register determine:

Timer or counter operation.
Enable or disable interrupts.
Positive or negative edge triggering.
Immediate or external start triggering.
Prescaler division ratio.
Reload time constant or not.
Reset or not.

The Prescaler

The prescaler counter is used only when the channel is being used as a **timer**. That is, the input to the channel comes from the system crystal clock. The prescaler is programmed by a bit sent to the channel control register to divide the incoming frequency by either 16 or 256.

The output pulse train from the prescaler is used as the clock to the **down counter**. In effect this further divides the system clock rate by another factor which depends upon the number loaded into the **time constant** register. Every time the **down counter** is decremented to 0, the zero count (ZC/TO) output is pulsed **high**.

An interrupt may also be generated. When the **down counter** reaches 0 it is automatically reloaded with the value from the **time constant** register.

The Time Constant Register

The time constant register is an 8-bit register used in both counter and timer mode. It must be programmed by the CPU just after the channel con-

trol register and contains a number between 1 and 256. This register loads its time constant into the **down counter** when the CTC is first initialised and then subsequently, whenever the counter is decremented to 0.

The Down Counter

This is an eight-bit register loaded initially and on reaching 0 from the time constant register. The counter is decremented by each external clock pulse in **counter** mode, or by the output pulse from the **prescaler** in **timer** mode. The CPU can read the contents of this register at any time by performing a normal port read operation.

Channels 0, 1 and 2 have zero count/time out (ZC/TO) pins which generate a positive pulse when the counter reaches 0. Channel 3 does not have this pin due to lack of space on the chip and is, therefore, restricted to the use of an interrupt to signal its zero count condition.

CTC Operation – Counter Mode

In this mode, the CTC counts externally applied pulses. These are applied to the **CLK/TRG** input of the required channel.

During initialisation of the channel control register, the mode of operation must be set to **counter**, and the positive or negative edge of the trigger pulse selected to be the one that decrements the **down counter**.

Initially, the **down counter** is loaded with the number from the **time constant** register. As **trigger pulses** are applied to the **CLK/TRG** input this number is gradually decremented. Eventually it reaches zero and the ZC/TO pin goes **high**. If interrupts have been **enabled**, an interrupt is simultaneously generated and this may be serviced by the CPU. In addition, the **down counter** is **reloaded** from the **time constant** register and counting continues with the arrival of the next trigger pulse.

For example, if a system was required which generated a pulse for every 100 trigger pulses, the **time constant** register would simply be loaded with 100. Then, for every 100 trigger pulses, the ZC/TO pin would pulse once. In addition, the CPU could be interrupted and this could take appropriate action for each 100 counts.

CTC Operation – Timer Mode

In **timer** mode, the CTC counts the system **clock** pulses. Because the clock frequency is very high, a **prescaler** is used to divide it down to a more useful frequency. Thus the prescaler can be made to divide by 16 or 256 depending upon the programming of the **channel control** register.

The clock pulses, divided in frequency by the **prescaler** value are then used to decrement the **down counter** in the same way as in the **counter** mode. Initially, the **down counter** is loaded from the **time constant** register and when it is eventually decremented to zero, the ZC/TO output goes **high**. This may also generate an interrupt.

The precise period of the pulses from the ZC/TO pin are given by:

$$t \times p \times \text{TC} = \text{pulse period}$$

where t is the system clock period
           p is the prescaler value (16 or 256)
and   TC is the **time constant** value.

Timing may be programmed to begin automatically as soon as the **time constant** word is loaded, or it may be started by an external trigger input on the **CLK/TRG** input.

Further details of the programming of the CTC are contained in the accompanying book, '*Practical Exercises in Microelectronics*'.

Summary

The main points covered in this chapter are:

- Input/output operations can be synchronised by using either a polling technique or an interrupt system.
- Interrupts are useful for irregular inputs since they are initiated by the peripheral device.
- Each peripheral port generally requires its own interrupt service routine.
- The CPU must be able to be interrupted and then continue its program as though nothing had happened.
- The stack is generally required to save the contents of the CPU registers at the beginning of the interrupt service routine. These are returned at the end of the routine.
- Interrupts may be non-maskable which means that they cannot be prevented by the software.
- Maskable interrupts may be non-vectored or vectored.
- Non-vectored interrupts cause a branch to a fixed location to find the service routine, whereas vectored interrupts supply the address via the interrupting port.
- Multiple interrupts require a priority determining system.
- The counter timer chip is capable of performing complex counting or timing operations to relieve the processor of a time consuming task.
- The CTC must be correctly initialised before it will function.

Questions

5.1 Briefly describe the circumstances in which it would be preferable to use a polling algorithm to an interrupt driven system.

5.2 What is the main reason for having an interrupt capability in a microprocessor system?

5.3 Explain why it is necessary to save the contents of the registers on the stack during an interrupt service routine.

5.4 Why is it necessary to have an enable interrupt instruction in an interrupt service routine?

5.5 What is the main difference between a maskable and non-maskable interrupt?

5.6 List the sequence of instructions that must be used to initialise a Z80 to expect and enable a non-maskable, non-vectored interrupt.

5.7 What is the difference between a vectored and a non-vectored interrupt?

5.8 How many **restart** instructions are there in the Z80 instruction set, and what are the addresses which they **call**?

5.9 The **initialisation** routine for a system with mode 2 interrupts can be divided into two parts, one to initialise each PIO and one to initialise the CPU.
  Write down the sequence of instructions that would be required to initialise the CPU for Mode 2 interrupts.

5.10 The service routine start address table contains the following entries:

> 1820 1B
> 1821 19
> 1822 20
> 1823 1A

Write down:
(i)   The contents of the CPU I register.
(ii) The interrupt vectors for the two ports in the system.
(iii) The start addresses of the two service routines.

5.11 Describe the advantages of a daisy chain priority system compared with that described by *Figure 5.3* earlier in the book (page 120).

5.12 List the three interrupt modes of the Z80 CPU and describe their functions.

5.13 It is required to use the Z80 in interrupt mode 1. Write a suitable CPU initialisation program and indicate where the service routine would be located.

5.14 Explain the following terms:
(i) Non-maskable interrupt.
(ii) Maskable, vectored interrupt.

5.15 How would the program be modified if the interrupt service routine start address table were located at address 1910 hex?

5.16 What would happen if the EI instruction is removed from the interrupt service routine?

5.17 How could the program be modified so that an interrupt is generated only when both bit 0 and 1 of the input port are at logic 0?

5.18 What are the two functions of the **CLK/TRG** input of the CTC?

5.19 What is the maximum time period for the output pulses in the **timer** mode, if the system clock is 1.79 MHz?

5.20 Why do only three CTC channels have a ZC/TO pin?

5.21 A CTC has a clock input of 2 MHz and it is required to generate pulses at 1 ms intervals. What values would need to be programmed into the **prescaler** and **time constant** register?

5.22 A Z80 CTC must be initialised to generate a precise pulse train with intervals between pulses of 1 ms on its channel 1 ZC/TO pin. This must be started by an external pulse. The clock runs at 2 MHz. Write a suitable initialisation program showing any calculations necessary.

5.23 Briefly explain how the interrupt priorities are assigned to the four channels of a CTC.

5.24 If the four channels of a CTC are connected together so that the ZC/TO pin of one channel is connected to the CLK/TRG pin of the next, and they are suitably programmed, what is the longest time interval that could be generated? Assume the system clock has a 500 ns period.

5.25 What is the shortest time interval that can be generated by the CTC?

SIX

Assembly language programming applications

When you have completed this chapter you should be able to:

1. Understand how to use a microcomputer system to write, assemble, run and debug programs involving assignment, selection and iteration.

2. Understand the need to convert data from one code type to another.

3. Explain the common microcomputer techniques for analogue-to-digital conversion software.

6.1 SOFTWARE TECHNIQUES

Some of the basic techniques employed in microcomputer programming have been examined in this and the accompanying books in the Microelectronics series. These include the use of loops in programs, time delays, and counting methods. Previous chapters have also introduced the techniques of initialisation required for both programmable input/output devices and counter timer chips. Some simple examples of the use of interrupts have also been included to show how this method may be used for irregular inputs of information from external devices. These techniques have all been relatively straightforward,

involving programs with only a few instructions. However, they are the building blocks of future programs which may be much more complex.

Designing computer software is very similar in some ways to designing hardware which, when the function of certain chips is known, becomes a matter of joining together the various building blocks required. There is a similar need to understand the basic techniques when designing software that will be reliable and also straightforward.

This chapter introduces some further techniques that may be used in many future programs. The first involves the conversion of data from one coded form to another. Microcomputers manipulate data and have to convert it from one code type to another in order to satisfy the demand of, for example, serial data links, binary outputs and digital display systems. All of these may require the same information coded in a different way.

Analogue data must frequently be manipulated by a computer, and therefore inputs in analogue form must be converted to digital format before they can be processed. The techniques required to do this may be implemented in hardware or in software, or a mixture of both. Two possible software techniques are examined in detail, which can either provide a very fast solution to conversion problem, or provide a solution with the minimum of programming.

6.2 CODE CONVERSION

Data in an 8-bit microcomputer system is stored in memory as 8-bit bytes, in binary code. However, the 8 bits may represent numbers in a variety of different formats, letters, instructions for the computer itself. Letters and numbers may be represented either directly in **binary** code, in **binary coded decimal**, in **ASCII** code, or in **7-segment** code. Under certain circumstances other types of code may also be used, such as the **scan** code generated when a keyboard is encoded, so that each code represents a specific key and therefore indirectly a letter or number. Depending upon the type of code used, either 4, 8 or more bits may be required to specify each letter or number.

Binary code may be used to represent numbers in a microcomputer. With 8 bits in the binary code, numbers between 0 and 255 may be directly represented by 1 byte. Binary code is by far the most common way of storing numbers since instructions exist which allow the binary code to be manipulated mathematically or logically to perform various operations required in computer programs.

Figure 6.1 shows the number 127 as it would be represented in binary code in a computer. It occupies 1 byte.

127

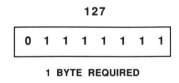

1 BYTE REQUIRED

Figure 6.1 127 in binary

Binary coded decimal is another common code used to represent numbers, particularly where a display is required in decimal format. The Z80 microcomputer contains a decimal adjust accumulator instruction (DAA), which allows numbers to be manipulated in a limited fashion using binary coded decimal format (BCD) (*Figure 6.2*). Using this code, each character in the number is represented by a 4-bit binary number, thus allowing two such numbers to be coded within the 8 bits of a normal byte. The example below shows the number 1, 2, 7 represented in 2 bytes although in fact only 12 bits are actually required. Each 8-bit byte is used to store 2 BCD characters, it is sometimes known as '**packed BCD format**'.

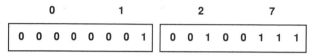

2 BYTES REQUIRED

Figure 6.2 127 in binary coded decimal

ASCII code (American Standard Code for Information Interchange) is the standard code used when data is transmitted between computers and peripherals. All of the letters, both upper case and lower case, the numbers, most punctuation marks and computer peripheral control characters have an ASCII code. This is a 7-bit code so that effectively one byte is required for each character. The ASCII code is shown in *Figure 6.3*. Very often when data is transmitted between computers, the bytes of data are separated into ASCII characters with each character representing 4 bits. This allows computer programs as well as alphanumeric data to be encoded in ASCII format for transmission between machines. However, it means that the data must be converted from binary to ASCII code in the transmitting machine, and then from ASCII back to binary in the receiving machine.

Following the example above, each of the numeric characters is preceded by a 3 to convert it to ASCII code, as shown in *Figure 6.3*. Therefore the number 1, 2, 7 would require 3 bytes to convert it to ASCII code (*Figure 6.4*).

7-segment code depends upon the hardware configuration of the machine which employs the 7-segment display. This means that there is not a fixed 7-segment code, but each one is machine specific, and depends upon which bit of the output port is connected to each segment of the 7-segment display. For example, if the bits of the

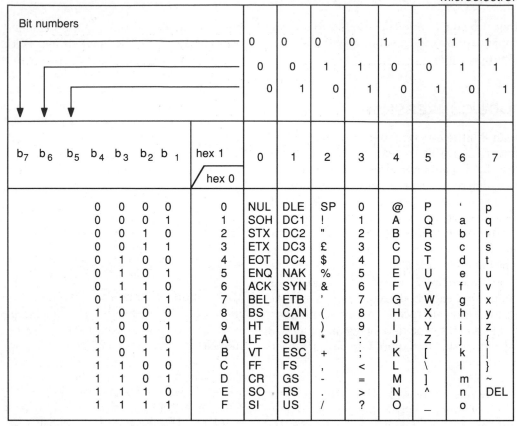

Bit numbers (b7 b6 b5)				0	0	0	0	1	1	1	1
				0	0	1	1	0	0	1	1
				0	1	0	1	0	1	0	1
b7 b6 b5 b4 b3 b2 b1	hex 1 / hex 0			0	1	2	3	4	5	6	7
0 0 0 0	0			NUL	DLE	SP	0	@	P	`	p
0 0 0 1	1			SOH	DC1	!	1	A	Q	a	q
0 0 1 0	2			STX	DC2	"	2	B	R	b	r
0 0 1 1	3			ETX	DC3	£	3	C	S	c	s
0 1 0 0	4			EOT	DC4	$	4	D	T	d	t
0 1 0 1	5			ENQ	NAK	%	5	E	U	e	u
0 1 1 0	6			ACK	SYN	&	6	F	V	f	v
0 1 1 1	7			BEL	ETB	'	7	G	W	g	x
1 0 0 0	8			BS	CAN	(	8	H	X	h	y
1 0 0 1	9			HT	EM	)	9	I	Y	i	z
1 0 1 0	A			LF	SUB	*	:	J	Z	j	{
1 0 1 1	B			VT	ESC	+	;	K	[	k		
1 1 0 0	C			FF	FS	,	<	L	\	l	}
1 1 0 1	D			CR	GS	-	=	M	]	m	~
1 1 1 0	E			SO	RS	.	>	N	^	n	DEL
1 1 1 1	F			SI	US	/	?	O	_	o	

Figure 6.3   ASCII code table

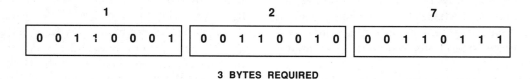

1	2	7
0 0 1 1 0 0 0 1	0 0 1 1 0 0 1 0	0 0 1 1 0 1 1 1

3 BYTES REQUIRED

Figure 6.4   127 in ASCII

port are connected to segments a,b,c,d,e,f,g and the decimal point, starting with bit 7 then the display codes would be those shown in *Figure 6.5* (overleaf).

Normally a logic '0' is required to illuminate a segment, and each segment is lettered in a standard pattern as shown in *Figure 6.6*.

6.3 CONVERSION METHODS

The process of converting data from one coded form to another depends largely upon the application. There are two basic methods:

(a) A **mathematical** method.
(b) A **look-up** table.

Mathematical methods tend to be used when there is a relationship between the two codes that holds true for all the data to be converted. The mathematical relationship clearly depends upon the two codes which are used and therefore a special program must be devised for each pair of codes to be converted. In addition separate programs will be required for code conversion from code A to B and code B to A since the process is not reversible. However, once a program is defined, it will work on all the input codes, no matter

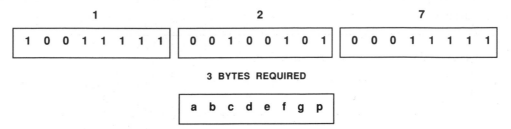

3 BYTES REQUIRED

a	b	c	d	e	f	g	p

Figure 6.5   127 in 7-segment code. Assume 7-segment connections and logic 0 to illuminate segment

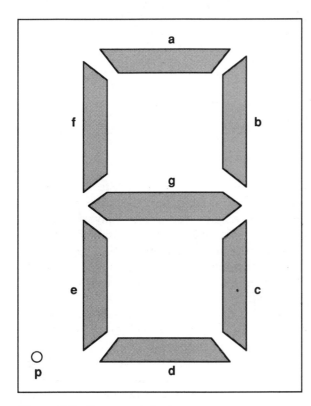

Figure 6.6   7-segment display

how many there may be. Generally, code conversion will always be done at the same speed, irrespective of the input data. Typical examples of code conversions that can be performed using a mathematical method are binary to BCD conversion and BCD to ASCII conversion.

The **look-up table method** of code conversion is more applicable to codes that bear no mathematical relationship to one another. A table must be created in memory that contains all the possible values of the input data, together with their corresponding converted code, generally stored

in a related memory address. The conversion software consists of a program that searches the input table for the code to be converted, and then uses its position in the table to determine the new code. Frequently, the old code and the new code are stored in adjacent memory addresses which make the processing very easy.

The main advantage of the look-up table method, is that one program can be used to perform the code conversion irrespective of the codes to be translated. All that changes for different codes are the input and output data tables. However, if a large number of values must be converted, then the size of the look-up table will become very significant. This means that it will take different lengths of time to find different values in the table, with those at the end requiring a longer conversion time than those at the start of the table. Often this is not a significant disadvantage, especially if the tables are relatively short. Typical examples of codes that could be converted using a look-up table are BCD to 7-segment code, or ASCII to 7-segment code.

One feature that both code conversion techniques have in common is that the input data and output data is normally held in specific memory addresses known as a 'buffer'. The conversion process then takes the data from the input buffer, performs the conversion process and then returns the new value to the output buffer. The number of bytes required for each buffer will depend largely upon the code that is being translated.

Case Study 1 –
Hexadecimal to ASCII Conversion

Hexadecimal numbers are used to represent 4 bits of a binary number, and therefore any byte can be

Table 6.1 Hexadecimal and ASCII codes

Binary	Hexadecimal	ASCII	ASCII (binary)
0 0 0 0	0	30	0 0 1 1 0 0 0 0
0 0 0 1	1	31	0 0 1 1 0 0 0 1
0 0 1 0	2	32	0 0 1 1 0 0 1 0
0 0 1 1	3	33	0 0 1 1 0 0 1 1
0 1 0 0	4	34	0 0 1 1 0 1 0 0
0 1 0 1	5	35	0 0 1 1 0 1 0 1
0 1 1 0	6	36	0 0 1 1 0 1 1 0
0 1 1 1	7	37	0 0 1 1 0 1 1 1
1 0 0 0	8	38	0 0 1 1 1 0 0 0
1 0 0 1	9	39	0 0 1 1 1 0 0 1
1 0 1 0	A	41	0 1 0 0 0 0 0 1
1 0 1 1	B	42	0 1 0 0 0 0 1 0
1 1 0 0	C	43	0 1 0 0 0 0 1 1
1 1 0 1	D	44	0 1 0 0 0 1 0 0
1 1 1 0	E	45	0 1 0 0 0 1 0 1
1 1 1 1	F	46	0 1 0 0 0 1 1 0

represented by two hexadecimal characters. When hexadecimal or binary numbers must be transmitted, they are frequently converted to ASCII code first. This means that each hexadecimal character becomes one ASCII code and one complete byte requires two ASCII characters.

Fortunately the process of converting from hexadecimal to ASCII is relatively simple and can easily be accomplished with a simple mathematical routine.

The codes that need to be converted are shown in *Table 6.1*.

To convert the numbers from 0 to 9 into ASCII code, they are simply preceded with 3. However,

translating the hex characters A to F requires a little more manipulation. In addition each hexadecimal character represents only 4 bits and therefore each byte of input data requires two ASCII characters to be produced. The requirements for the input and output buffer are therefore shown below in *Figure 6.7*, where each input byte requires two output bytes.

Although the diagram shows only two input bytes and four output bytes, the process will convert any number of bytes.

The conversion process is very simple. If the input code is less than or equal to 09, then 30 is added to it to convert it to the new code. If the

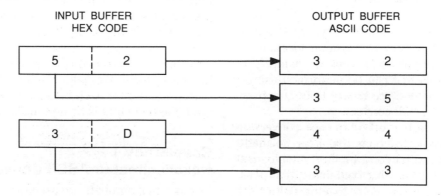

INPUT BUFFER
HEX CODE

OUTPUT BUFFER
ASCII CODE

Figure 6.7 Hexadecimal to ASCII conversion buffers

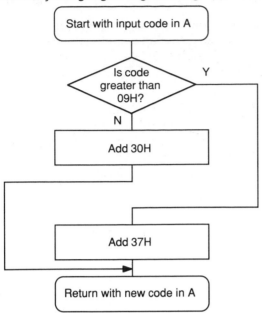

Figure 6.8 Hex to ASCII conversion subroutine

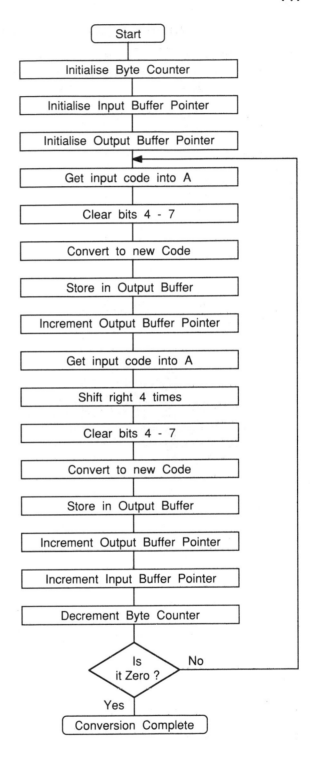

code is greater than 09 then 37 is added to it. Since this operation would normally be repeated a number of times in any conversion process, it is convenient to write it as a subroutine, as shown in *Figure 6.8*.

The subroutine assumes that the input code is in bits 0 to 3 of the accumulator, and it produces the result also in the accumulator.

This makes it relatively easy for the programmer to arrange for the input data to be in the same format for each conversion operation, and the output data can easily be stored from the accumulator. The complete conversion flow chart is shown in *Figure 6.9*. This is arranged to perform the conversion irrespective of the number of bytes, by setting the byte counter at the beginning of the routine to the number required. The next two instructions then initialise the pointers to the input and output buffer. The main loop of the program converts one complete input byte into two ASCII characters. It does this first by converting the lower 4 bits, and storing them in one of the output buffer locations, then it retrieves the upper 4 bits, shifts them right four times so that they appear in the lower 4 bits of the accumulator and then performs the same conversion routine. This process is repeated for a number of bytes in the input buffer.

Figure 6.9 Complete hex to ASCII flow chart

Case Study 2 – ASCII to 7-Segment Code Conversion

Converting from any type of code to 7-segment code, is best achieved by using a look-up table since the 7-segment code has no fixed mathematical relationship with any other code.

This can be seen by examining the codes shown in *Table 6.2*, which assumes that a logic '0' is used to illuminate each segment, and the connections between the display device and the output are as shown, in simple alphabetical order. Many systems employ completely different connections, and therefore the actual 7-segment code will be very different, although the conversion principle is exactly the same.

Table 6.2 ASCII and 7-segment codes

Character	ASCII code	7-Segment code	Display segments
0	30	03	a b c d e f
1	31	9F	b c
2	32	25	a b d e g
3	33	0D	a b c d g
4	34	99	b c f g
5	35	49	a c d f g
6	36	41	a c d e f g
7	37	1F	a b c
8	38	01	a b c d e f g
9	39	09	a b c d f g
A	41	11	a b c e f g
B	42	C1	c d e f g
C	43	63	a d e f
D	44	85	b c d e g
E	45	61	a d e f g
F	46	71	a e f g

7-segment code assumes segment connections are
a b c d e f g p and a logic 0 illuminates each one.

The conversion technique using a look-up table has two elements:

(a) The table containing the input and output codes.
(b) A program that searches the table for input codes and then looks up the output code.

It is up to the programmer how the look-up table is constructed, but a common method is to have adjacent bytes with input and output code respectively. These codes will be converted into a look-up table as shown in *Figure 6.10*.

30	ASCII 0
03	7 - SEGMENT 0
31	ASCII 1
9F	7 - SEGMENT 1
32	ASCII 2
⋮	
61	7 - SEGMENT E
46	ASCII F
71	7 - SEGMENT F

Figure 6.10 ASCII and 7-segment code look-up table

The heart of the software that looks up the values of the input code in the table is a subroutine shown in *Figure 6.11*. This starts with initialisation instructions for the length of the table, or the number of possible values and then its location in memory. Each value is compared in turn with the code for which a conversion must be found. If a match is found the table pointer is incremented and the value returned from the next memory address. If the initial value is not equal to the unknown code the table pointer is incremented twice to the next ASCII value, the counter is decremented, and the process repeated until a match is found. If no match is found in the entire table then the input code is assumed to be an error and a dummy value must be returned from the subroutine.

Managing the data in and out of the conversion routine is a main program which simply takes a value from the ASCII input buffer, converts it and stores it in the 7-segment buffer. The way this is achieved is a standard method using a byte counter for the number of bytes that must be

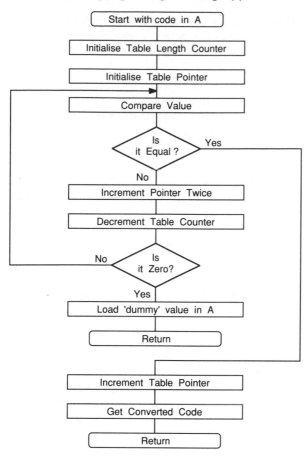

Figure 6.11 Look-up table subroutine (convert byte)

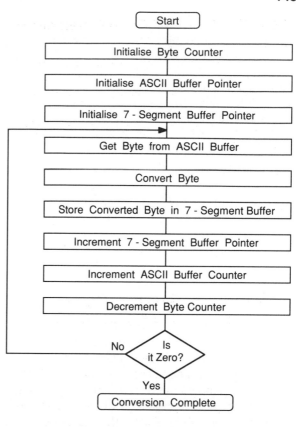

Figure 6.12 ASCII to 7-segment conversion flow chart

converted and incrementing each buffer pointer so that they remain in step (*Figure 6.12*).

The main program and conversion subroutine shown above can clearly be used for many different types of code conversion simply by changing the data in the look-up table. As such it has many useful functions. However, it is not the only software that could be used to perform this operation, and other methods exist, particularly where the input data has a specific numeric range.

6.4 ITERATION

Iteration is a technique that is used in many branches of computing and mathematics. It involves 'making an estimate of a correct solution, then adjusting the estimate in steps until it is as close as possible to the required result'.

Generally with iteration techniques, the first estimate should be as close as possible to the correct solution, so that the number of additional attempts is kept relatively small. This gives a 'rapidly converging' solution.

A typical mathematical example of the use of an iteration technique, is the method that can be used to calculate the square root. The method can be described simply as follows. First make a guess at the square root, then square it and see how close it is to the original number. Then adjust the original guess either positive or negative to attempt to bring a next estimate closer to the actual square root. Continue this process until the square root calculated is as close as possible to the actual value which when squared gives the original number.

A typical example in the field of microelectronics is the method used to convert an analogue value to a digital number. This process takes place within every analogue-to-digital converter which is used to allow analogue inputs to be read and processed by digital systems. It is therefore of particular significance, and the method of iteration determines the overall operating speed of the analogue-to-digital conversion.

There are many techniques for performing analogue-to-digital conversion, but the two that are described in the following case studies are known as:

(a) The **ramp** method.
(b) The **successive approximation** method.

Both have their advantages and disadvantages, and are therefore covered in some detail.

Case Study 3 – Analogue-to-digital Conversion by the Ramp Method

The method is based on the use of an electronic comparator chip, with very high sensitivity, such that its output will be in either a logic '1' or a logic '0' state, depending upon the relative values of the voltages at its two inputs. A difference of more than a few millivolts will force the device output to be in either one state or the other.

By applying an unknown voltage to one input, and a **ramp** of known voltage to the other input, the output can be made to change state when the ramp voltage reaches the value of the unknown voltage. In this way the value of the unknown voltage can be determined. This is shown in *Figure 6.13*.

Clearly, this technique cannot work unless the value of the ramp voltage is known at every point in time, so that at the time in *Figure 6.13*, when the comparator changes state, the ramp voltage must be known.

One way of achieving this is to generate the ramp from a digital-to-analogue converter, driven either by a computer output port, which sends increasing binary numbers to the output, or from a binary counter which performs the same operation in hardware. If the whole operation is to be driven by computer, the comparator output must also be fed back to an input port, so that the computer knows when the operation is complete. This is shown in *Figure 6.14*.

The output provided by the computer cannot be a smooth ramp in the same way that an analogue circuit could produce one. It must be a series of steps (*Figure 6.15*), with each voltage dependent upon the binary output of the port. With an 8 bit port the output will be a series of up to 255 steps. The height of each step will be dependent upon the total voltage range which the digital-to-analogue converter covers. For a typical 5 V output, each step will be approximately 20 mV. Therefore when the comparator changes state because its step ramp voltage has exceeded the unknown analogue input, it must be slightly higher than the analogue voltage by up to 20 mV. In practice this is not generally a problem since it is better than 0.5 per cent accuracy on the full-scale value.

The main advantage of using this stepped ramp method of analogue-to-digital conversion, is that the software required to generate the ramp is very simple, and the technique is therefore very easy to

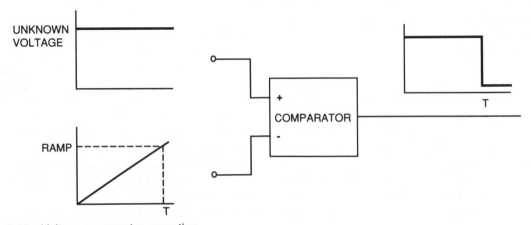

Figure 6.13   Voltage comparator operation

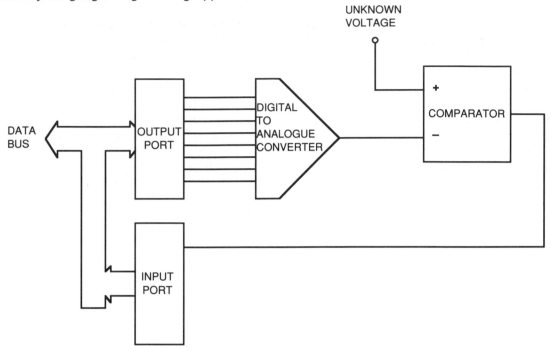

Figure 6.14 Computer controlled ramp ADC

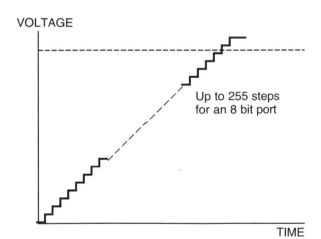

Figure 6.15 Stepped ramp output from DAC

understand and to troubleshoot should it go wrong. If the duration of each step is relatively short, say 5 or 10 ms, depending upon the settling time of the digital-to-analogue converter, then the maximum conversion time can be relatively short, perhaps only 2.5 ms.

However, the fact that the ramp must rise to the correct voltage from a zero start, means that the conversion time is longer for higher voltages, and shorter for lower ones. In some systems this variable conversion time may be a problem. Also for high-speed conversions, the time of 2.5 ms could be unacceptable, since it only allows about 400 samples per second.

Some of these problems may be overcome by using the same hardware, but a different conversion technique known as **successive approximation**.

Case Study 4 – Analogue-to-digital Conversion Using Successive Approximation

Although the ramp method is successful at performing analogue-to-digital conversion using a relatively simple technique, its major disadvantage is that the time taken to perform the conversion process varies depending upon the value of the unknown voltage. For low voltages the process is very rapid, whereas for higher voltages it takes much longer. In some systems this would present a serious problem, so if a method can be found that will allow the process to take an equal time whatever the voltage this would be a significant improvement.

The method that achieves this is known as **successive approximation**. It is much more complex in terms of the software, but can be achieved using exactly the same hardware, which is that shown in *Figure 6.14*.

In this method the unknown voltage is compared with a voltage generated by the computer, in exactly the same way as the ramp method. However, this time, instead of starting with a zero voltage, the computer generates a voltage which starts at half the maximum. This is achieved by setting bit 7 of the digital output number to a logic '1'. If the unknown analogue voltage is higher than half the maximum voltage, the bit in bit position 7 is retained and the software continues by setting bit 6. However, if the unknown analogue voltage is less than half the maximum value the bit in bit position 7 is reset to zero and the process continues with bit 6. It can be explained in more detail by looking at an example of the bits which would be set and retained or set then reset by examining *Figure 6.16*.

If the unknown voltage was a value as shown in *Figure 6.16*, the process would take place as follows. First bit 7 is set and the computer-generated voltage compared with the unknown. Since the computer output is lower than the unknown voltage, the bit is retained in bit 7 as a logic '1'. Bit 6 is then set and this adds to the voltage already generated as a result of bit 7 so that a higher output voltage is produced. This also is lower than the unknown, so bit 6 is also retained. Bit 5 is now

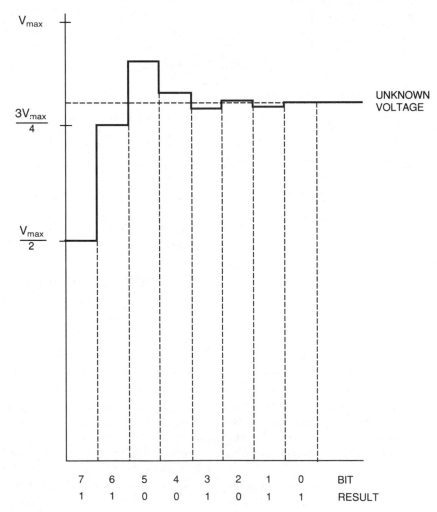

Figure 6.16   Successive approximation

set but this generates a computer output that is higher than the unknown so this bit is reset to a logic '0'. This same process is then applied to bit 4 and when this is set the computer-generated voltage is still higher than the unknown so this bit also is reset. Bit 3 together with bits 7 and 6 produce a lower voltage than the unknown, so bit 3 is retained. The process is repeated with bits 2, 1 and 0, until the resulting computer-generated voltage is as close as possible to the unknown. In this case the final bit pattern is 1 1 0 0 1 0 1 1.

The whole process takes only eight samples and is therefore very rapid. Each sample need only be stable for as long as it takes for the comparator to react which can be within a few microseconds. A typical conversion time with this process can be as short as 100 μs.

This process can be built into the logic of a chip, so that the whole process takes place in hardware, but in this case, a computer controls the operation with software. Naturally this software is more complex than that required to control the ramp method, and *Figure 6.17* shows a typical flow chart.

Three main registers have been used in the flow chart. One is the bit register which holds a single 1 in the bit being tested at any moment. The second is the successive approximation register which holds the resulting data, and the third is a counter which simply counts the number of bits to be tested. It would be possible to dispense with the counter and simply wait for the single bit in the bit register to be rotated out altogether, but the counter has been used for the sake of clarity.

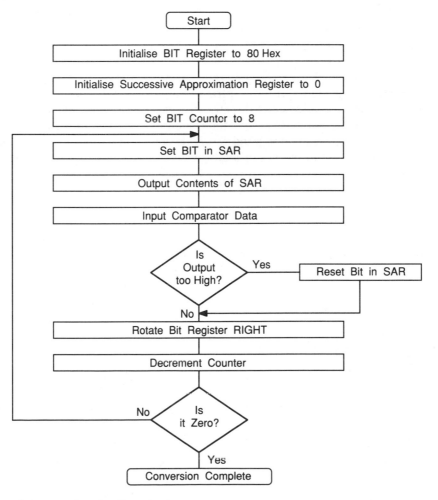

Figure 6.17 Successive approximation flow chart

The main advantages of this method, as have been indicated above, are its speed of operation and the fact that whatever the unknown voltage, the conversion time remains the same.

Both this method and the ramp conversion method are accurate to within half the least significant bit, and they both suffer from the same problems which relate to noise on the input voltage. If any noise is present, or if the signal varies during the conversion time, then errors can result which will reduce the accuracy of the conversion. Professional systems would normally employ 'sample and hold' circuits on the analogue inputs.

6.5 SUBROUTINES AND SERVICE ROUTINES

One of the advantages of the use of assembly language, over machine code, is that it encourages 'top-down' program design. This means that a flow chart can be produced in which general statements are made about the steps in the program, and these can often be translated directly into subroutines. The effect of this is that the overall program design can be analysed and decided before the detail of the subroutines or interrupt service routines is required.

A good example of this is shown in *Figure 6.18*, which forms part of a clock routine.

The statement in the flow-chart boxes state the required action at each step, but contain no detail on how each step will be performed. However, this flow chart may be translated directly into the

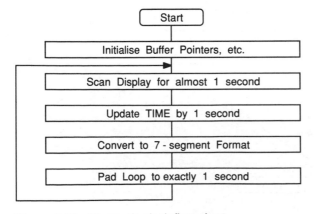

Figure 6.18   Electronic clock flow chart

```
; CLOCK ROUTINE
          ORG 1800H
MAIN:     CALL INIT      ; INITIALISE BUFFERS
LOOP:     CALL SCAN      ; SCAN DISPLAY
          CALL UPDATE    ; UPDATE TIME
          CALL 7SEG      ; CONVERT FORMAT
          CALL PAD       ; PAD TO 1 SEC
          JR LOOP        ; REPEAT
```

Figure 6.19   Clock routine main program

main program of the routine by assuming each step will be a subroutine. This is shown in *Figure 6.19*.

The use of assembly language also encourages the use of named routines, since at the program production stage, the addresses are not generally known. This in turn leads to the possibility of using subroutines in more than one program and hence building up libraries that will ease the production of software at a later date. Routine libraries tend to be a feature of some of the modern high-level languages such as C or Modula 2, but can also be effectively used at assembly language level. However, it must be noted that whenever library routines are used the documentation that accompanies them must be thorough, indicating the exact effect of the routine, together with any changes in the register contents or other relevant information.

The techniques of assembly language programming described in this and previous chapters are examined in much more detail in the book *Practical Exercises in Microelectronics* which forms part of this series. It is recommended that those practical exercises are studied alongside the theory, particularly for the more complex aspects of microcomputer software development.

Summary

The main points covered in this chapter are:

- Good software is the key to successful computer applications, and it is important that a good programming technique is used from the outset. The best approach for assembly language programming is to think in terms of the overall function of the software, then work down to the detail. This is known as a **top-down** approach.

- Data conversion can occupy a significant amount of microprocessor time. It can be performed either by a mathematical method or by the use of a look-up table.

- Analogue-to-digital conversion can be performed by using iteration, successively approximating to the required solution.

- Subroutines and the use of interrupts encourage top-down program design.

Questions

6.1 Give two reasons for using BCD code in computer systems.

6.2 Explain briefly what is meant by a **software buffer**.

6.3 Briefly describe one method of transferring binary data from one computer to another.

6.4 Explain briefly how the ramp method conversion process and the successive approximation method would deal with an analogue voltage higher than the maximum which could be generated by the computer.

6.5 What is the main disadvantages of the ramp method of analogue-to-digital conversion software?

6.6 Why would data conversions involving 7-segment code normally require a **look-up** table?

SEVEN

Practical printed circuits

When you have finished this chapter, you should be able to:

1. Relate logic circuit diagrams to printed circuit board (PCB) layout.
2. Explain the effects of inductance, capacitance and resistance associated with PCBs on the high-speed digital signals.
3. Understand how buffer elements can prevent 'ringing' in bus lines.
4. Explain how decoupling elements can help to eliminate cross-talk.
5. Explain the methods of PCB production.

7.1 PRINTED CIRCUIT FUNDAMENTALS

Before printed circuit boards (PCBs) were invented, there was no recognised method of interconnecting electronic components to form circuits. The circuits tended to be designed with valves, and often the resistors and capacitors etc. were simply attached to the valve bases by whatever means was thought practical. Large components such as transformers were simply connected to the equipment chassis and leads were attached in a haphazard fashion between components. As the circuit sophistication increased, small components were mounted on tag strips and these allowed a slightly improved layout to be achieved.

Component sizes tended to be much larger than they are today, which meant that manual wiring of the relatively simple circuits was straightforward. As component sizes reduced, and circuit complexity increased, the old methods of interconnection were no longer of any value, and the printed circuit was invented. This provided both the mechanical means of support for the components, and the necessary electrical interconnections which reduced the wiring considerably. Printed circuits, like all other components, have developed considerably in their own right over the years. The modern PCB has almost as much bearing upon the performance of a system as do the other major components in it.

The universal application of PCBs to electronic design has forced upon component manufacturers design rules that make the task of circuit production relatively easy. For example, because printed circuits are most easily produced when the holes are on a fixed grid pattern, component manufacturers tend to produce components with the appropriate lead pitches so that they are very easily mounted. Until the advent of surface mount technology, most components were designed with a lead pitch which was a multiple of 0.1 inches (*Figure 7.1*). This means that almost universally, integrated circuits, resistors, capacitors, transistors, connectors and switches etc. can be placed upon a 0.1 inch grid, which makes the design of the PCB very simple.

Each electronic component has a pattern of pads associated with its lead positions. These pads represent the areas that will be used for the solder to connect the component to the copper

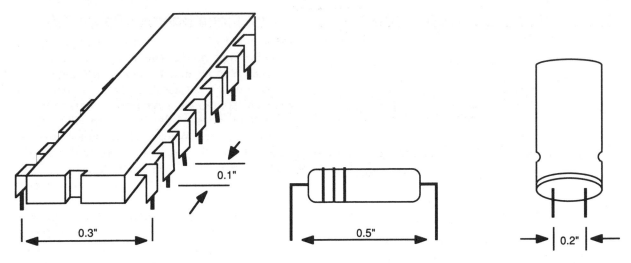

Figure 7.1 Typical component lead spacing

track on the circuit board. In addition the pads have a marker for the hole position so that the leads position can be punched or drilled through the circuit board. *Figure 7.2* shows some typical examples.

Most professional system designers use artwork that is twice full size, with the pads and other artwork scaled accordingly. This allows a greater degree of accuracy but it requires photographic reduction before it can be used to produce the final PCB. For simple applications on an amateur basis, most artwork is designed at normal size so that it can be transferred directly to the PCB and the final product produced very easily.

For modern circuits with surface mounted components, which have no leads going through the board, the pin spacing may be considerably smaller than 0.1 inches. Therefore, although the same design principles apply, a typical grid would have a 0.025 inch spacing to allow each lead to register with one of the grid lines.

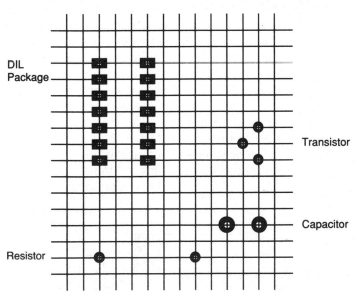

Figure 7.2 Typical component pads

7.2 TYPES OF PRINTED CIRCUIT BOARD

There are a number of different types of PCB commonly in use, each of which has its own application. Some of the most common types are indicated below.

Single-Sided Boards (*Figure 7.3*)

Single-sided boards, as the name suggests, have copper on one side of the board with the components all mounted on the other side. They are mostly used in the entertainment electronics industry where manufacturing costs must be kept to a minimum. However, in some industrial electronics where cost is important and where the circuitry is not too complex single-sided boards are frequently used. Their main disadvantage is that with some circuit configurations it is not possible to arrange all the tracks without having to include wire 'jumpers' to make all the connections. If a circuit design and layout require a large number of jumpers, then a double-sided board should be considered.

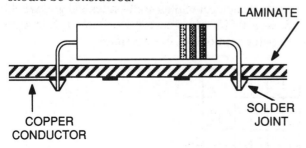

Figure 7.3   Single-sided PCB

Single-sided boards are by far the most popular for non-professional applications, since they are so easy to manufacture in a workshop or laboratory. They are also very easy to manufacture using 1 : 1 artwork.

Double-Sided Boards (*Figure 7.4*)

Double-sided PCBs can be made with or without plated-through holes. The normal double-sided boards are simply an extension of the single-sided boards in which the number of conductors or the density of components required necessitates the use of both sides of the boards for the copper interconnections. Where connections are required between the sides of the board either components are soldered on both sides, or special pins are placed through the board and soldered on both sides.

Double-sided boards tend to be more expensive than single sided, but they are much less expensive than those with plated-through holes. They generally require hand soldering on one side of the board to ensure that the necessary connections are properly made.

Double-Sided Boards with Plated-Through Holes (*Figure 7.5*)

Double-sided boards with plated-through holes are very common in industrial electronics and particularly in computer circuits. This is because they allow very high component densities, and circuit interconnections can be achieved very easily. Every hole effectively connects one side of the board to the other, so that conductors can be routed from one side to the other to avoid obstacles. This feature is of particular importance when automatic component layout or track routing is used, and special holes are drilled in the board known as **vias** which make these connections between sides.

The price of plated-through hole boards is higher than normal double-sided boards, and in particular every additional hole adds a small amount to the price of the board. If the cost is an important consideration, therefore, many design-

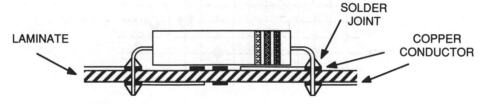

Figure 7.4   Double-sided PCB

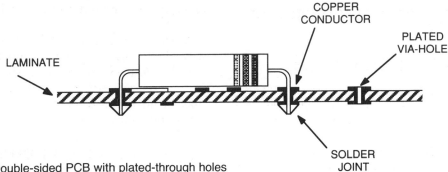

Figure 7.5   Double-sided PCB with plated-through holes

ers may prefer to use a standard double-sided board with the addition of a few interconnections between the sides.

Multi-Layer Boards (*Figure 7.6*)

With the advent of plated-through hole technology, a considerable increase in component packing density has been made possible. However, with the VLSI devices now commonly available, so many interconnections are required that in many instances even double-sided boards are very tightly packed. This gives rise to unpredictable design problems such as noise, crosstalk, stray capacitance, and inadequate decoupling between parallel signal lines. The only solution to these problems was the introduction of a multi-layer board, with 4, 6 or even 10 layers.

In a basic 4-layer board, two of the layers are reserved for the earth and power connections with the other two being the component interconnections. This arrangement overcomes many of the problems because an earth rail is effectively placed close to each signal conductor. This minimises cross-talk by shielding each conductor and providing a high distributed capacitance to decouple the power supply rails.

The multi-layer board consists effectively of a number of thin PCBs stacked together and adhesively joined to form one rigid board. This will be slightly thicker than the normal double-sided board. Electrical connections between the different layers are achieved with plated-through holes. Where the inside connectors are joined to the plated-through hole, the conductor width is increased to slightly more than the hole diameter. The hole drilling process will therefore expose the bare copper of the conductor around the hole for the plating-through process.

Multi-layer boards are generally only used when absolutely necessary, although they account for perhaps 30 per cent of the PCB market. They are naturally very expensive, but can be cost effective since it is possible to achieve up to 50 per cent higher packing density than would be possible with a double-sided board. In addition their per-

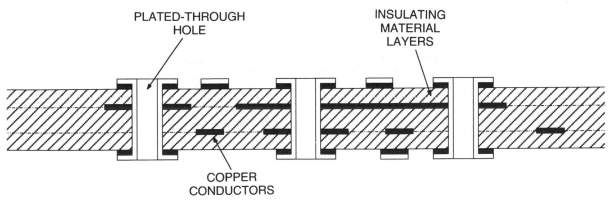

Figure 7.6   Multi-layer PCB

formance improvement may be sufficient to justify the additional cost.

Flexible Printed Circuit Boards

Flexible PCBs have many applications in the electronics industry where both their flexibility and weight give them an economic advantage over traditional assembly methods. They may simply function as a connection harness between other boards, or may carry components in their own right. They account for about 10 per cent of the PCB market, and are used to best advantage where they are the complete interconnection system within a device. They can be manufactured in either single, double, or multi-layer configurations, using a range of base materials and conductors. Their only real disadvantage lies in their ability to handle high frequencies.

Some of the more common materials used for the PCB include epoxy resin/glass matt, polyester foil, polyimide foil, and Teflon foil. Each of these has slightly different characteristics in terms of its flexibility, tear resistance, flame retardant capability, etc. Special considerations have to be made when designing with flexible boards, such as the shape of the solder pads, and the avoidance of using solder connections in areas that will be bent. These areas are particularly prone to conductor breakage if solder pads are used in them. If they are to be subject to a large number of bending operations, single-sided flexible boards tend to perform better (*Figure 7.7*). In addition, the larger the bending radius, the less likely any fracture will be.

Boards are generally covered with a film to increase the physical strength, and this must have holes left in it for the solder pads. Some boards also are provided with stiffeners which are attached to the back of the boards in the component mounting areas.

7.3 PRINTED CIRCUIT BOARD PRODUCTION

PCBs can be produced by a variety of means, depending upon the complexity of the circuit involved. In very simple cases, where a single-sided board is required, the simplest method involves sketching the pattern of copper tracks and component pads directly onto a copper-clad board with an etch-resist pen. Once dried, the board with the sketched copper pattern can be directly etched and a completed board produced in a very short time. However, the quality of the product leaves a lot to be desired, and is only suitable for one-off operations.

All boards of a professional or semi-professional nature start with a detailed circuit diagram, from which precision artwork is produced.

The importance of a precise and detailed circuit diagram cannot be overemphasised, since any errors in the circuit will automatically be reflected in an incorrect design which will not have a chance of working correctly. A PCB should be

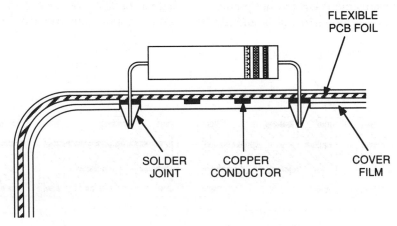

Figure 7.7   Flexible single-sided PCB

produced only when the prototype circuit has been thoroughly tested.

From the working circuit diagram and a knowledge of the size of the required PCB the artwork for the printed circuit can be produced either by manual means or with the aid of a computer. In either case, although the pattern of copper pads and tracks is the most important, at least four and possibly five separate diagrams need to be produced. These are:

(a) Component overlay.
(b) Copper track and pads (component side).
(c) Copper track and pads (copper side).
(d) Solder resist pattern.
(e) Drilling detail.

The copper track and pads on the component side of the board are required only when a double-sided PCB is to be produced.

Manual Artwork Production

PCB artwork is produced by manual methods in many industries, where a need for computer-aided techniques is not justified by the expense.

Imagine that a small PCB is required to carry a driver transistor and light emitting diode to be used as an indicator in a piece of electronic equipment. *Figure 7.8* shows the various pieces of artwork required to produce the circuit from the

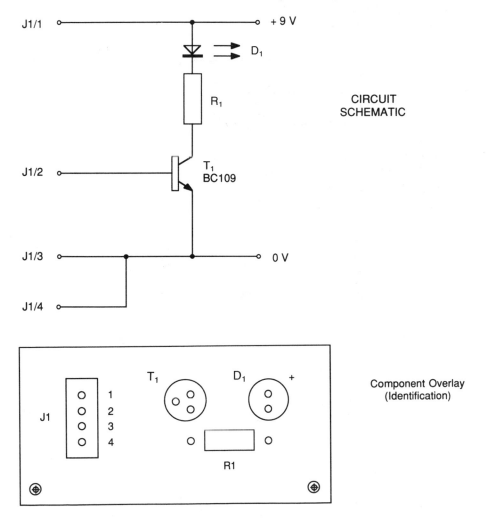

Figure 7.8(a)–(b) PCB artwork

schematic shown. The various stages in its production would be as follows.

Any design must start with an accurate circuit diagram, and as shown in the schematic in *Figure 7.8*, all components need to be numbered, and any connectors need to be clearly identified. Care must be taken to ensure that all component numbers are unique, so that there is no confusion when the board is finally produced. The schematic is also likely to include other values such as voltages which may not appear on the final board pattern.

Generally, the artwork designer needs to work with three diagrams simultaneously. One for the component overlay which indicates where the components will be placed, and either one or two indicating the copper pads and tracks, depending upon whether single-sided or double-sided boards are being produced. In the case of the example shown in *Figure 7.8*, only a single-sided board is required and therefore the designer would work on two layers simultaneously.

The most common design method is to place a 0.1 inch precision grid pattern on a drawing board or on top of a 'light box', and cover this with the required number of sheets of drafting film. Tracing paper can be used but this is less dimensionally stable than professional drafting film and is therefore not generally recommended.

The first step is to mark the board outline on each of the layers that are being worked upon. The designer then places the component mount-

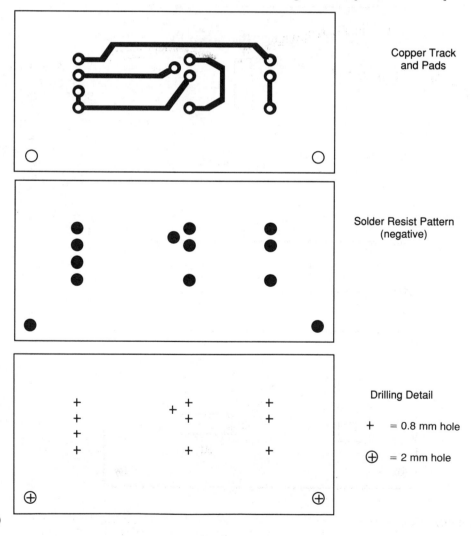

Copper Track and Pads

Solder Resist Pattern (negative)

Drilling Detail

+   = 0.8 mm hole

⊕   = 2 mm hole

Figure 7.8(b)

ing pads and their overall outlines on the appropriate layers to achieve a satisfactory component layout. Simultaneously, the mounting pads are joined together with thin precision tapes to represent the copper tracks. The patterns representing the pads and the interconnecting tapes can normally be supplied from electronic component suppliers. Mounting pad patterns are normally supplied on tear-off strips or rolls, and have an adhesive backing so that they stick to the drafting film. Alternatively they may be supplied as rub-on transfers, although these seem to produce a less satisfactory overall result.

Some designers place all the components on the board first, and then go back and make the necessary interconnections. However, the majority work simultaneously on the component placement and interconnections so that by working in the direction of signal flow, signal paths can be kept short and critical components can be placed close together. There may be a number of specialist design considerations which must be taken into account with different types of circuit, particularly with high frequencies or high-speed operation and these are dealt with later in this chapter.

Once the complete layout and copper track pattern has been produced, the professional circuit designer will also need to produce a pattern for the solder resist, and the drilling details. Solder resist is the green-coloured material that generally coats PCBs. However, there are holes in its pattern so that the copper pads are exposed and can be used for soldering. The solder resist prevents solder from sticking to other parts of the board and producing faults due to short circuits, etc.

Where boards are being produced by a PCB manufacturer, drilling details are also required. This allows the manufacturer to drill all the holes of the correct size, particularly where mounting holes or other large holes are required. At this point, the designer may need to refer to the actual components to ensure that the holes he requests are large enough to accommodate the leads on the components to be used.

Once the artwork has been produced, the PCB can be manufactured, but before this takes place another step may be required if the artwork has been produced at twice full size. Most professional artwork would be photographically reduced to a 1 : 1 scale before the manufacturing process. In addition, if multiple boards were required the artwork may be placed a number of times on the same negative so that a number of boards can be produced simultaneously. This is achieved in a step-and-repeat photographic process.

Computer-Aided Artwork Production

An alternative method of producing the PCB artwork is to use a computer-aided design package. These are now readily available at relatively low cost and can produce professional quality artwork with a minimum of effort. Most packages now provide *auto-routing* which means that the designer only needs to produce the circuit schematic, and the layout diagram. The computer automatically makes all the connections. The steps in the process are illustrated in the diagrams of *Figure 7.9* (pages 158–61).

The steps in the process are as follows. The PCB designer uses the computer-aided design package to first draw a detailed circuit diagram. Most PCB packages contain a library of components and these can be extracted and placed on the computer screen wherever required. The computer prompts the user for pin numbers where required and component identification numbers. The designer builds up the circuit making the necessary interconnections between components by drawing lines on the computer screen.

When the design is finished, it is saved on disk, and the computer produces a **net-list**. This **net-list** is a list of all the components, together with their pin numbers and the interconnections between the pins. Some systems also produce a complete component list.

The next step in the process is to produce a layout diagram. Once the board outline has been drawn, to the correct size, the components required are extracted from the component layout library and placed on the board in the required position. Some systems prompt the designer to use the same components which have been identified as part of the schematic circuit diagram. Others leave it to the designer to remember to include all the components. Once all the components have been placed, the job of the designer

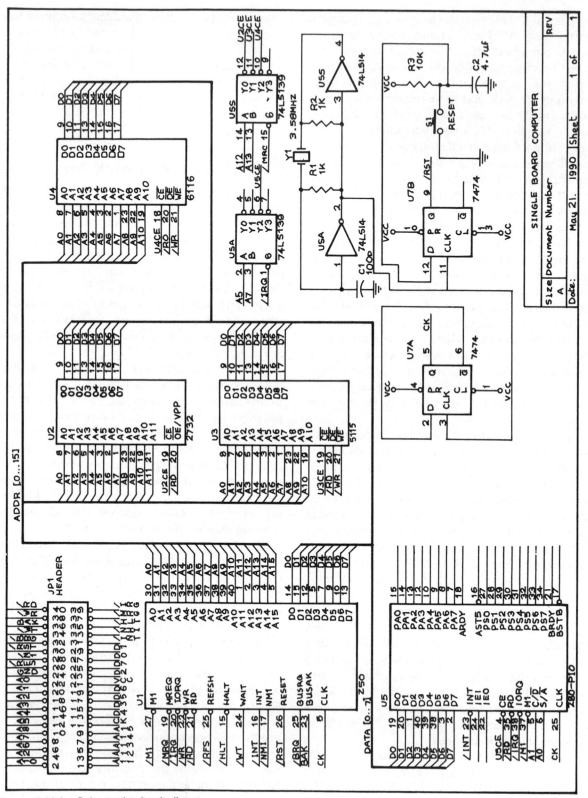

Figure 7.9(a)   Schematic circuit diagram

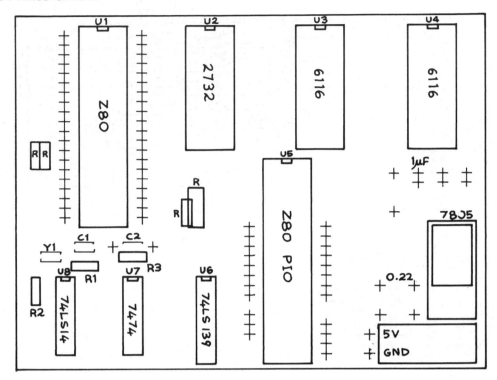

Figure 7.9(b) Component layout

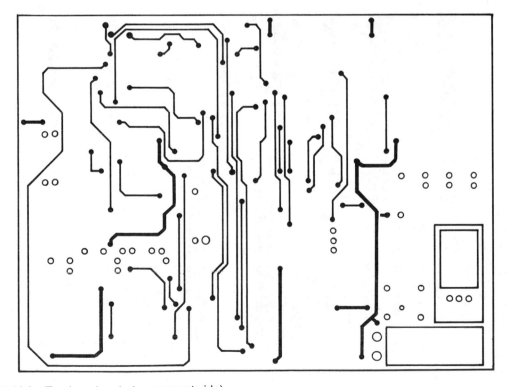

Figure 7.9(c) Track and pads (component side)

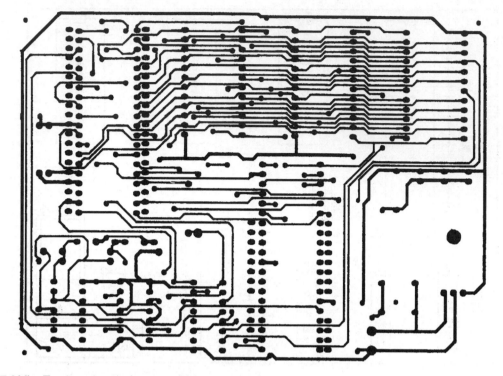

Figure 7.9(d) Track and pads (copper side)

Figure 7.9(e) Solder resist pattern

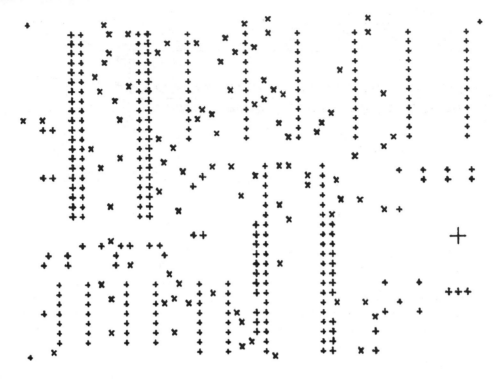

Figure 7.9(f)   Drilling pattern

is then finished, and the computer completes all the required interconnections with its auto-router.

The **auto-routing** program takes the component layout diagram and the positions of the components, together with the **net-list** produced from the schematic diagram, and it attempts to connect all the pins with the copper track. If a double-sided board has been requested, the computer may need to generate extra holes so that tracks can be routed from one side of the board to the other and so complete the interconnections. If some attempts fail, the more sophisticated auto-routers will rip up the tracks already laid and re-route them in an attempt to route every track. It is generally possible to define the parameters within which an auto-router will work. For example, it may be necessary to specify a minimum track width, or the maximum number of turns which a track may make. Some tracks may need to be wider than others, and these can also be specified, sometimes before the computer routes other tracks. If an auto-router fails to complete a board, the designer is left to finish the more difficult connections manually.

The circuit diagram and component layout shown in *Figure 7.9(a)* and *7.9(b)* have been converted to the track layout shown in *Figure 7.9(c)* and *7.9(d)*. This represents a relatively rough version of the final artwork, which is done for test purposes on paper rather than on the drafting film used for the final artwork.

In addition, the computer generates both the solder resist pattern and the drilling pattern from the information it contains. These are shown in *Figure 7.9(e)* and *7.9(f)*. In this particular case, the computer has been asked to generate a double-sided printed circuit board with plated-through holes.

Board Manufacture

Once the PCB artwork has been produced, the board itself may be manufactured in a number of ways depending upon the type of board required.

A simple single-sided PCB can be produced from 1:1 artwork in a laboratory. This is achieved as follows:

(a) Use a copper-clad laminated circuit board,

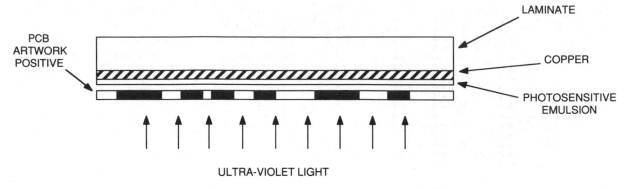

PCB
ARTWORK
POSITIVE

LAMINATE

COPPER

PHOTOSENSITIVE
EMULSION

ULTRA-VIOLET LIGHT

Figure 7.10   Producing the etch-resist pattern

which has a coating of photo-sensitive emulsion on the copper surface. Place the required
artwork on the photographic emulsion, and
expose it to ultraviolet light for a few minutes
(*Figure 7.10*). This exposure is normally done
upside down in an ultraviolet light box. The
exposure to ultraviolet light chemically
changes the areas where copper is not
required.

(b) The exposed areas where copper is not
required are removed by dipping the board
for a few moments in a dilute alkaline solution. This removes the photo-sensitive emulsion, leaving only the exposed pattern of the
required conductors and pads.

(c) The board is then etched in a ferric chloride
solution to remove the copper that is not
required, leaving only the pads and the tracks
that were defined by the artwork. Care must
be taken at all stages in the process to ensure
that pattern is accurate, and in particular during the etching process that neither over- or
under-etching takes place.

(d) Once the copper tracks are left on the board it
can then be drilled with the required holes for
component mounting.

A similar process can be carried out if a double-
sided board is required by exposing both sides of
the board to ultraviolet light as described above,
using the appropriate artwork for each side.
However, great care must be taken to align the
patterns correctly, otherwise when holes are
drilled they will not meet the appropriate pads on
the opposite side of the board. The process is

summarised in *Figure 7.11*, which shows the professional method of double-sided PCB production.
The only difference between this and the previous
technique is that a different photo-sensitive emulsion is used which requires a negative film master
of the artwork. This is because the emulsion har-

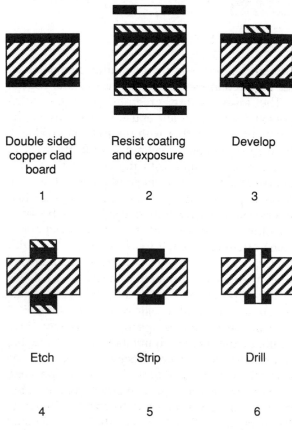

Double sided copper clad board	Resist coating and exposure	Develop
1	2	3
Etch	Strip	Drill
4	5	6

Figure 7.11   Double-sided print and etch process

dens under the exposure to ultraviolet light. The unexposed areas are then easily removed in a simple washing process once the pattern has been developed.

Boards With Plated-Through Holes

A number of processes exist for the production of double-sided boards with plated-through holes, such as the panel-plating process, the tenting process, and the pattern-plating process. However, of these the pattern-plating process is the most widely used. Each stage in the pattern-plating process is shown in *Figure 7.12*.

The process begins with a double-sided copper covered board. The first step is to drill all the required holes.

To make the holes electrically conducting, they need to be covered with a layer of copper, and this is done in two stages. The first stage is known as **electroless copper plating**. It deposits a very thin layer of copper over the entire surface of the board by a chemical process. The board is first activated by coating it with a thin film of palladium and it is then immersed in a copper sulphate solution. This, together with a reducing agent, deposits copper on the board surface, and on the inside surface of the holes. The layer of copper is very thin, and must be strengthened by electrolytic plating.

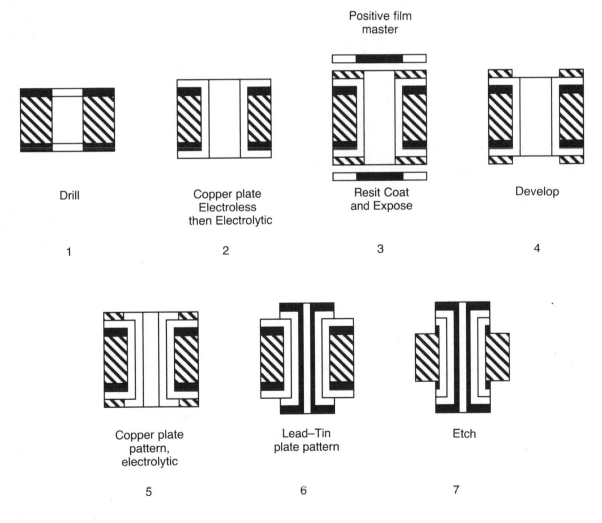

Figure 7.12 Plated-through holes – pattern-plating process

The **electrolytic plating** process involves using the board as an electrode in a copper sulphate bath and passing a d.c. current through the solution. Copper is deposited on the board surface, a few μm thick.

The board is then coated with a photo-sensitive emulsion, and exposed using a positive film master of the required artwork. When the unexposed emulsion is removed, this leaves the copper areas required exposed for the next process.

The main thickness of copper on the tracks and pads is now built up in a further electrolytic plating process, typically to a thickness of 35 μm.

Before the photo-sensitive emulsion is removed, the further plating process takes place with a lead–tin mixture, which not only improves the solder ability of the board, but also acts as an etch-resist for the next process.

Finally, all the unwanted copper from the surface of the board is removed in an etch bath, with a suitable chemical such as alkaline ammonia. Those areas that are covered with the lead–tin are not etched in this process and therefore remain as the track, pads, and plated-through holes of the board as designed.

The etching of printed circuit boards is a large subject in its own right, since it must be carefully controlled in large-scale production processes. The concentration of the solution, the etch time, the temperature etc. must be carefully controlled to avoid under-etching board and so leaving copper on the surface, while also avoiding over-etching and creating open circuit tracks. Note that in the production of boards with plated-through holes, by the process outlined above, the etching chemical cannot be ferric chloride or cupric chloride, which are the most common etching solutions since they both attack the lead–tin etch-resist used in the process. However, these chemicals do tend to be used widely with normal single-sided or double-sided PCB production.

7.4 PRINTED CIRCUIT BOARD PARAMETERS

In simple low-frequency applications, the PCB can be seen as a method of interconnecting the circuit components. However, the board is a component in its own right, and as such has various characteristics. These generally only become significant at high frequencies, or with high currents or voltages.

Digital systems use very high frequency pulses, which are likely to be affected by the board parameters, and also surges of current along the very thin tracks may frequently cause problems. Typically, the capacitance between tracks causes cross-talk between the signals they carry. Also mismatching of the circuit impedances to the transmission line characteristic impedance presented by the tracks on the board can cause ringing, etc. The serious board designer must be aware of the current carrying capability of the PCB and also take care to minimise the risks involved where high frequencies or high voltages are concerned. This is done very simply by observing some well-documented design rules.

The main features that a board designer must take into consideration, therefore, are its resistance, capacitance, inductance, and the transmission line effects that the layout creates.

Resistance

The copper conductor tracks on a PCB have a finite resistance, which introduces a voltage drop proportional to the current flowing in that conductor. The value of the resistance R is given by the formula:

$$R = \frac{\rho \times l}{A} \, \Omega$$

where ρ = resistivity ($\Omega \times 10^{-6}$)
l = length of track (cm)
A = cross sectional area (cm^2)

The standard thickness of copper foil is assumed to be 35 μm without any plating. This gives a resistance value of about 5 mΩ for a 1 mm wide conductor which is 1 cm long. The resistivity is taken to be 1.7241 at 20 °C.

As a typical example, consider the resistance of one of the bus lines on a microprocessor board, which may have a width of 0.5 mm, and be about 20 cm long. This would have a resistance of:

$$R = \frac{1.7241 \times 10^{-6} \times 20}{35 \times 10^{-4} \times 0.05} \ \Omega$$

$$R = 0.2 \ \Omega \ \text{approx.}$$

The value of 0.2 Ω may seem insignificant, but it is large enough to have to be taken into consideration, particularly as it rises with temperature.

For example, if the PCB is inside an enclosure which rises in temperature to perhaps 85 °C, when the outside temperature is 20 °C, the resistance would rise by approximately 25 per cent.

In practice, most of the conductors in a microprocessor based circuit carry small currents, for which the conductor resistance can be practically ignored. However, the supply and ground lines generally ought to be much wider than the signal lines to improve the high frequency performance and reduce high voltage spikes which appear on them due to high current.

Capacitance

Two conductors separated by a non-conductor, can be considered to form a capacitor. This applies to PCBs when conductors on opposite sides of the board are involved, or when adjacent conductors on the same side of the board are involved.

The most obvious capacitive effect occurs in double-sided boards (*Figure 7.13*).

An approximation for the capacitance between conductors on opposite sides of a PCB can be made using the formula:

$$C = \frac{0.886 \times \epsilon_r \times A}{d} \ \text{pF}$$

where A = overlapping area (cm^2)
ϵ_r = relative dielectric constant
d = dielectric thickness (mm)

Typical values for the relative dielectric constant of laminates vary between about 4.2 and 8.0. Therefore typical capacitance values can easily be calculated as follows.

For example, consider a PCB with relative dielectric constant of 5.4. It has two tracks on opposite sides of the board, each 2 mm wide, and separated by a board thickness of 1.6 mm. Their common length is 250 mm.

Using the formula the capacitance can be calculated as:

$$C = \frac{0.866 \times 5.4 \times 0.2 \times 25}{1.6} \ \text{pF}$$

$$C = 15 \ \text{pF}$$

A capacitance of 15 pF can be significant in coupling signals between adjacent tracks on a PCB, but it may also serve a useful purpose if it decouples the power supply rail to ground. This would be the case if the power rail ran on one side of the board immediately above the ground rail on the other.

Capacitance between adjacent conductors is rather more difficult to calculate (*Figure 7.14*). It depends upon the conductor width, their separation, and the thickness of the copper on the board.

A formula may be used to calculate the actual capacitance, but for practical purposes it depends more upon the conductor width than the separation, for separation values greater than 1 mm. If the relative dielectric constant is taken to be 5.4 for example, and the thickness of the conductor

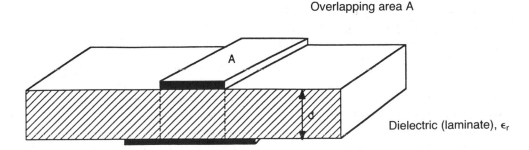

Overlapping area A

Dielectric (laminate), ϵ_r

Figure 7.13 Capacitance on a double-sided board

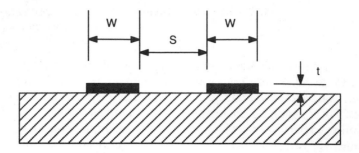

Figure 7.14   Capacitance between adjacent conductors

35 μm, then with 2 mm spacing and 2 mm wide conductors, the capacitance would be 0.4 pF per centimeter of parallel track. This would also be the value for 1 mm wide conductors separated by 1 mm. 1 mm wide conductors separated by 2 mm give about 0.3 pF/cm.

Consider a simple example. Two conductors run in parallel for 150 mm. They are 1 mm apart and each 1 mm wide. The total capacitance between them would be 0.4 pF/cm:

$$0.4 \times 15 = 6 \text{ pF}$$

Inductance

Two tracks on the same side of a PCB as shown in *Figure 7.14* also exhibit some inductance. Once again the formula to calculate the exact value of this inductance is very complex, but some typical values can be given which may be useful in practical situations. Generally the inductance decreases as the conductor width increases and it also increases as the spacing increases.

For two conductors each 1 mm wide, separated by 1 mm, the approximate value of inductance would be 8.5 nH/cm. If the spacing increases to 2 mm for the same conductors the inductance increases to 10 nH/cm. For conductors of 0.5 mm at a spacing of 2 mm the inductance would be 12 nH/cm.

Transmission Line Effects

Perhaps more important than the actual inductance exhibited by conductors on the PCB is the effect that the inductance and capacitance have in creating a transmission line from parallel tracks.

The major characteristic of a transmission line is its characteristic impedance Z_C, which is given by the formula:

$$Z_C = \sqrt{\frac{L}{C}} \ \Omega$$

where L =   inductance per unit length
       C =   capacitance per unit length

Using the values previously described, it can be calculated that the approximate characters for impedance for two parallel conductors 1 mm wide, separated by 1 mm, would be approximately 140 Ω. However, the actual value in any practical situation will vary considerably from this value and may be between 50 Ω and 500 Ω.

Signals sent along transmission lines suffer reflections whenever an impedance that does not match the transmission line impedance is encountered.

The reflection sends a pulse back along the conductors which is then re-reflected at the sending end, and this process continues until the reflect amplitude becomes very small. However, its effect is to modify the waveshape significantly, and in some cases disastrously.

Figure 7.15 (pages 167–8) shows the effect of various transmission line impedances. The diagrams clearly show a number of important points.

Since it is the receiving end voltage that must most closely resemble the input pulse, on both rising and falling edges, then the closest approximation is achieved if the value of Z_C is 150 Ω. Even so, the waveforms are not ideal and both the rising edge and falling edge exhibit a small overshoot. The important point is that the rise time is maintained.

The voltage at the sending end of the conductor

SENDING END 0 – 1 EDGE

$Z_c = 50\ \Omega$

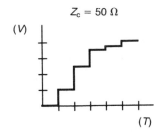

$Z_c = 100\ \Omega$

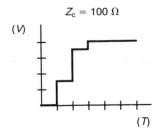

$Z_c = 150\ \Omega$

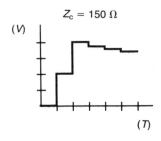

RECEIVING END 0 – 1 EDGE

$Z_c = 50\ \Omega$

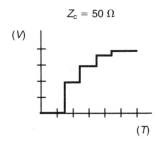

$Z_c = 100\ \Omega$

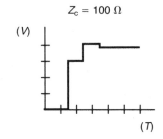

$Z_c = 150\ \Omega$

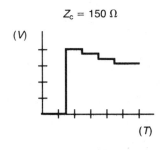

SENDING END 1 – 0 EDGE

$Z_c = 50\ \Omega$

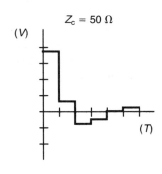

$Z_c = 100\ \Omega$

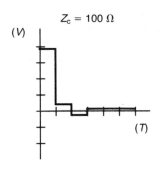

$Z_c = 150\ \Omega$

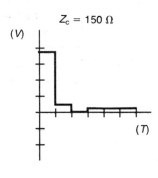

RECEIVING END 1 – 0 EDGE

$Z_c = 50\ \Omega$

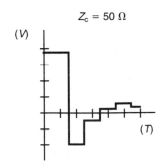

$Z_c = 100\ \Omega$

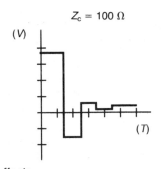

$Z_c = 150\ \Omega$

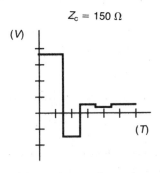

Figure 7.15(a)   Pulse transmission line effects

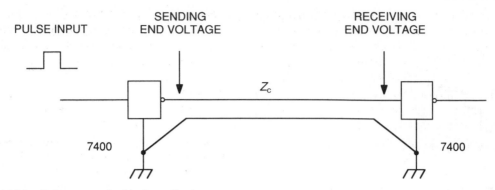

Figure 7.15(b)   Pulse transmission line effects

has a much slower rise time than that of the receiving end, particularly as the Z_C is reduced. It is therefore important that if another gate is connected to the system its input is connected close to the receiving end gate rather than directly to the output of the sending end gate.

The lower the line impedance, the worst the waveform appears to be. The receiving end rising edge is much slower, and the falling edge has a greater overshoot. This overshoot can lead to ringing in the bus lines. The diagrams show idealised square edged pulses, but in practice these would be rounded because of the frequency limitations of the devices involved, together with capacitance effects.

Cross-talk is another transmission line effect, which occurs when two signal conductors run parallel to each other on a PCB. Generally unless their parallel length exceeds about 20 cm for TTL then cross-talk is likely to be insignificant. However, in some cases it is unavoidable and a pulse on one conductor can introduce a spike or a train of pulses on the adjacent conductor (*Figure 7.16*).

Cross-talk effects are much more difficult to analyse than reflections, since they effectively involve two sets of transmission lines. There is an impedance between each signal line and ground, together with an impedance between the signal lines.

In general, the further apart the signal lines can be kept, and the closer they can be to the ground line the less the cross-talk will be. This means that in critical situations, the best method of eliminat-

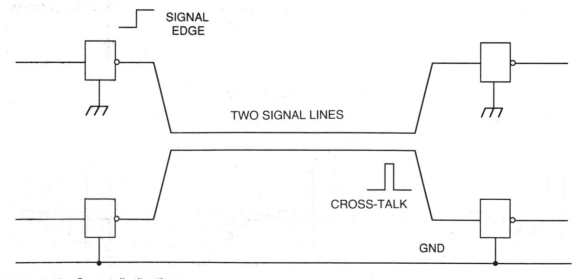

Figure 7.16   Cross-talk situation

ing cross-talk is to place a ground plane or ground track immediately below the signal conductors on the other side of the PCB. On a single-sided board it is best to place a ground conductor between the two signal lines which is virtually guaranteed to eliminate the possibility of any cross-talk. The presence of a ground line close to a signal line effectively decouples the signal by providing capacitance along its length.

7.5 PRACTICAL PCB DESIGN CONSIDERATIONS

Good circuit design and PCB layout are very important, particularly for high-frequency or high-speed digital circuits. By observing the basic rules for the type of logic devices being used, many of the potential problems can be overcome at the design stage, and reliable circuits produced. Designers must pay attention to the features such as:

(a) Track width and layout.
(b) Elimination of 'ringing' and the use of line drivers.
(c) Elimination of cross-talk.
(d) Elimination of ground and power supply noise.

Track Width and Layout

The width of tracks on a PCB should not be an arbitrary choice. Of particular importance are the widths of the supply and ground lines, since these have a significant effect upon the performance of the circuit and the elimination of a number of problems. The ground potential must be stable, and therefore in most circumstances the width of the ground conductor should be as large as possible, particularly for TTL circuits although this is less critical with CMOS designs.

For any circuit the fundamental rule is:

Ground width > supply width > signal width

However, for TTL circuits the following applies in particular:

Ground width > twice supply width
Supply width > twice signal width

The layout of ground and supply lines should be such that they are placed either directly opposite one another on each side of a double-sided board, or adjacent to one another on a single-sided board. This provides capacitance between the two and effectively decouples the supply rail.

The width of any signal lines on TTL circuits should be such that any transmission lines created have a characteristic impedance which is between 100 and 150 ohms. In practice this means that the conductor width should be $0.5 \times$ the board thickness, and therefore track widths between 0.5 and 1 mm give good results. Broad signal tracks that effectively create low-impedance lines should be avoided since they are highly susceptible to current spikes.

In MOS-based circuits, signal line widths are less critical, and good results are obtained with thinner conductors. This is because the MOS devices have much higher input impedances.

The design layout and routing of printed circuit tracks is largely a matter of common sense. For double-sided boards, the normal convention is to attempt to run tracks in one direction on one side of the board and at right angles to this on the other side of the board. This allows the number of additional holes to be minimised and produces a reasonably well distributed conductor pattern.

It is always best to use the shortest possible route for conductors, and attempt to start and end them on solder pads, rather than joining them to other conductors. Also sharp internal angles of less than 60° should be avoided. The minimum recommended spacing for tracks is normally the same as the conductor width, but wherever a more generous spacing can be used it is recommended that this is done. Not only does this reduce the interline capacitance, but it improves the reliability of the manufacturing process.

Eliminating 'Ringing'

'Ringing' is created in a circuit when the input and output impedances of the gates do not match the characteristic impedance of the transmission lines created by their interconnections. The problem is more complex than may at first be realised because the impedances of TTL and other gates are non-linear, which makes simple solutions very difficult. When lines between components are kept very short, the problem is largely eliminated, but in some circumstances long lines are unavoidable and therefore at high speeds action must be taken to reduce possible 'ringing'.

There are two possible solutions, one involving the addition of resistive components to the circuit, and the other the addition of buffers or line driver chips (*Figure 7.17*).

Some improvement can be made in the trailing edge response of TTL circuits by the addition of a resistor in series with the output of the driving gate. Typically this should be 10 ohms less than the characteristic impedance of the line. Resistors at the receiving end of the line may also improve

both rising and falling edge performance, but because they present a drain across the power supply they are not generally recommended. Probably the best solution where long lines are involved, or in particular where a PCB is connected to a bus system within the computer chassis, is the addition of special buffers or line driver chips to the circuit. These represent an additional expense, but for practical purposes they can improve the performance significantly. Their main function is simply to match the impedances of the lines to which they are connected, and therefore eliminate the reflections which would otherwise occur. In their most common configuration, line drivers and receivers are built into the same pack-

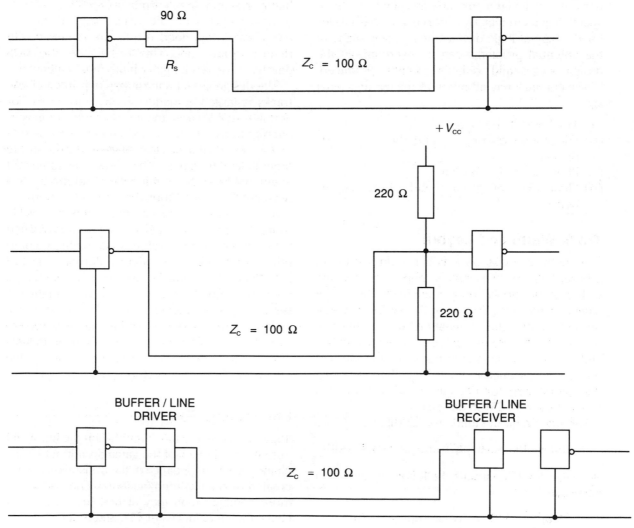

Figure 7.17   Reducing line 'ringing'

age as transceiver chips. This means that the devices may be used as either transmitter or receiver at either end of the line.

Eliminating Cross-talk

Cross-talk is more of a problem in TTL circuits than in CMOS or other MOS circuits, but with careful design it can easily be eliminated.

The most important consideration is to keep both conductors as far apart as possible, and to ensure that each is as close as possible to a ground line. Where two signals follow parallel tracks, it is also important to ensure that the signals are both travelling in the same direction. This means that the gate outputs both drive the same end of the pair of lines. If the opposite occurs, and the signals are travelling in opposite directions this can lead to a much more critical situation and a greater likelihood of cross-talk. In this circumstance, if it cannot be avoided it is best to place a ground track between the signal tracks.

Logic families with a higher noise immunity are generally less susceptible to interference from cross-talk, or other sources. This means that CMOS devices are probably the best in this respect, although of course they have other disadvantages such as their relatively slow speed.

Supply Line and Ground Noise

Perhaps one of the most serious problems with TTL and other digital circuits is the noise that comes from the supply or ground lines. During the transition of all gates, and in particular TTL, there is a current spike which flows through both of the output devices. For a TTL gate this may only last 5 ns, but can be as high as 20 mA. In addition, when an output changes state there may be another spike which is caused by the charge or discharge of the capacitance of the transmission line which is effectively connected to the gate output. This could contribute another 20 mA. Both of these current spikes must be carried by the same V_{CC} and ground lines, as shown in *Figure 7.18*.

The current of only 10 or 20 mA for 5 ns may seem insignificant. However, when a board is considered which may have 10 or 20 gates changing states simultaneously, this can lead to total currents from the supply of 200 mA to 400 mA. This will not only create a voltage drop due to the line resistance, but may also cause the power supply voltage to sag if it is incapable of delivering this level of current.

Fortunately there are a number of well-proven methods to overcome these problems.

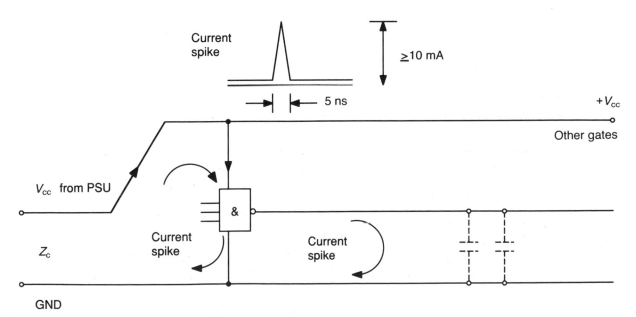

Figure 7.18 V_{CC} and ground-spikes

(a) Provide ceramic decoupling capacitors at a large number of points on the board. Typically this should be every two to three chips. Capacitor values of 10 nF for TTL and 5 nF for CMOS. It is important that the decoupling capacitors used are not electrolytic or have a wound construction, since these introduce an inductive element which is undesirable.

(b) Provide a low characteristic impedance between the V_{CC} and ground lines. Ideally, V_{CC} should be between 5 and 10 mm wide at least, situated close to or on the opposite side of the PCB of a double-sided PCB to a large ground line.

(c) Provide the ground with as large a copper surface as possible so that it has very low resistance and also provides some electro-magnetic screening. Much of the unused area on a PCB could be connected to ground with good effect.

(d) Avoid the use of the same ground lines for digital and analogue parts of the circuit. Separate ground connections should be taken to an individual point within the equipment.

A summary of the recommendations for PCB design, both for TTL and MOS circuits, are shown in *Table 7.1*. It also shows how to eliminate some of the undesirable effects which have been discussed.

Table 7.1 PCB design summary

Feature	TTL	MOS
Recommended Z_c between signals and ground	100–150 Ω	150–300 Ω
Recommended width of signal tracks	0.5 × board thickness	0.5 × board thickness
Signal to ground distance	As near as possible	Not nearby
Recommended V_{cc} width	10 × board thickness	3 × board thickness
Recommended Z_c between supply and ground	< 5 Ω	< 20 Ω
Recommended ground width	As wide as possible	Not too wide
Eliminating reflections and ringing	Use thin signal lines Use resistive matching Use line drivers	Use thin signal lines far from ground Use line drivers
Eliminating cross-talk	Use nearby ground	Not critical in general
Eliminating supply and ground spikes and noise	Use adequate decoupling Use a large ground plane	Use decoupling Ground plane not critical

Summary

The main points covered in this chapter have been:

- PCBs are not only used to interconnect circuit components, but must be considered as a component in their own right.
- PCBs may be either single sided, double sided, double sided with plated-through holes, multi-layer or flexible.
- PCB artwork is normally designed using a 0.1 inch grid pattern.
- Artwork may be produced either using a manual tape and transfer method or using a computer-aided design package with automatic track routing.
- Boards are produced by photographically transferring the required conductor pattern to the board and then etching away the copper that is not required.
- Plated-through hole double-sided boards are generally manufactured using a pattern plating process.
- Conductors on a PCB exhibit effects related to their resistance, capacitance, inductance, and transmission line characteristic impedance.
- Transmission line effects on PCBs give rise to 'ringing' and pulse overshoot.
- Line drivers or buffers may be used to eliminate 'ringing' in circuits.
- Cross-talk can be a problem between adjacent conductors which are not placed close enough to a ground plane.
- Power supply, noise and spikes may be eliminated with the use of decoupling capacitors at regular intervals, and an adequate ground plane.

Questions

7.1 Under what circumstances would a PCB designer choose 1 : 1 for the layout?

7.2 Briefly describe why double-sided plated-through hole boards are the norm in the digital electronics industry.

7.3 Compare the advantages and disadvantages of using a computer-aided design package for the production of PCB artwork.

7.4 Briefly explain the difference between electroless and electrolytic copper plating.

7.5 The track on a standard PCB is 0.5 mm wide and 12 cm long. Calculate its resistance.
    If its width is increased to 1 mm, how will this affect its resistance?

7.6 The relative dielectric constant of a PCB laminate is 5. Calculate the capacitance between two conductors on either side of the board, whose thickness is 1.6 mm, if they are 2 mm wide, and run opposite one another for 12 cm.

7.7 Briefly explain why 'ringing' occurs on some signal lines on a PCB?

7.8 A circuit board must be connected to another which will be about 50 cm away. Suggest a method of overcoming any potential problems with this arrangement.

7.9 A PCB with mainly TTL devices, appears to suffer from large voltage spikes on the supply lines. Suggest methods that may be used to overcome this problem.

Answers

CHAPTER 1

1.1 CMOS and TTL logic gates are fabricated using entirely different types of transistor. The CMOS gate uses field effect transistors of both the P channel and N channel types, while the TTL gate uses bipolar NPN transistors.

Their logic functions may be similar, but the CMOS gate will have much lower power dissipation, much higher fan-out, but operates at a slightly slower speed.

1.2 A circuit designer may prefer to use a standard 74 series chip in preference to a 74LS device, if the circuit to be designed must operate at high speed, and the power supply requirements are not par-

ticularly critical. Some 74 series chips may also be cheaper than their 74LS counterparts.

1.3 54LS244.

1.4 −225 mA.

1.5 0.4 V.

1.6 See *Figure A1.6* (below).

ROM 1	LOW	0000
ROM 1	HIGH	0FFF
ROM 2	LOW	1000
ROM 2	HIGH	1FFF
ROM 3	LOW	2000
ROM 3	HIGH	2FFF
ROM 4	LOW	3000
ROM 4	HIGH	3FFF
ROM 5	LOW	4000

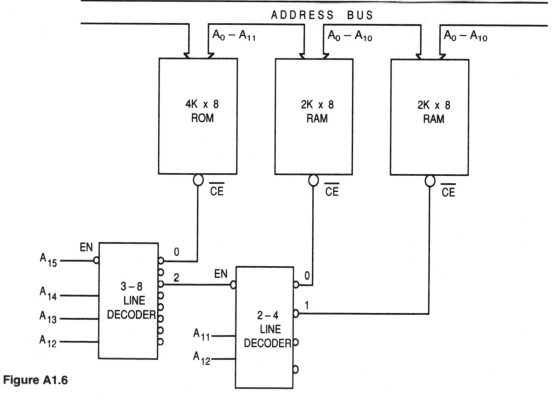

Figure A1.6

175

ROM 5	HIGH	4FFF
ROM 6	LOW	5000
ROM 6	HIGH	5FFF
ROM 7	LOW	6000
ROM 7	HIGH	6FFF
ROM 8	LOW	7000
ROM 8	HIGH	7FFF

1.8 The main advantage of partial address decoding is that the minimum number of decoder chips are used in a circuit that reduces the circuit cost. However, this can have a disadvantage that the memory is mirrored in a number of address ranges and this can lead to strange results if programs do not operate correctly. It also makes it very difficult to expand the system by adding additional memory without also adding additional decoder chips.

1.9 Linear address decoding is generally used only in very small microcomputer systems with one or two memory devices. This is because it leads to difficulties in organising programs to use the address spaces created by the linear decoding.

1.10 The advantages of using memory mapped input/output (I/O) are:
(a) The same instructions can be used to refer to either a memory location or a port address.
(b) It is often possible to use the same decoder circuits for both I/O devices and memory, thus reducing the circuit complexity.

1.11 This is shown in *Figure 1.12* (page 16).

1.12 Dynamic memory has the advantage that it has a very high storage capacity because of the high packing density within each chip. Therefore, with a given number of chips a large amount of data can be stored. It is also relatively cheap and has a high speed compared with some static devices.

Its main disadvantages are that it requires additional circuits to refresh the information every 2 ms, and that it often requires multiple power supply lines such as $+5$, -5 and $+12$ V.

1.13 The 74LS139 is a dual 2–4 line decoder chip with active low outputs and a single active low enable input for each decoder. Its other features are that it is a standard TTL device with the normal fan-in, fan-out and power supply requirements.

1.14 Some microprocessors have no input/output control line and are therefore forced to use memory mapped input/output.

CHAPTER 2

2.1 If more than one device can output data to the data bus at any moment, the possibility exists for a bus conflict. This means that two devices can attempt to force the data bus into incompatible logic states. This could result in two transistors being placed across the power supply rails while they are both conducting, which would lead to a very high current being drawn, and both devices are likely to be destroyed.

2.2 All devices connected to the data bus must have the common feature of a tri-state output. This allows them to activate a high impedance state when the output pin is not required to transmit or receive data.

2.3 The chip **enable** signal is used to indicate how a device will operate, whether it will read or write data or whether it is being treated as a memory device or an input/output device by the CPU.

2.4 An **EPROM** would require the $\overline{\text{RD}}$ and $\overline{\text{MREQ}}$ signals to be active simultaneously to correctly **enable** it.

2.5 A **bidirectional buffer** is used only when the data bus of a system is relatively large, i.e. the system contain a large number of memory or input/output chips. In small systems the output capability of the CPU will generally be sufficient to drive the devices connected to the data bus.

2.6 The main CPU control group signals for a Z80 processor are:
$\overline{\text{HLT}}$ – HALT
$\overline{\text{RESET}}$ – Reset
$\overline{\text{WAIT}}$ – Wait
$\overline{\text{INT}}$ – Interrupt request
NMI – Non-maskable interrupt

2.7 The Z80 requires a special clock circuit if it is to be used at high speed so that the rising edge of the waveform is fast enough to activate the microprocessor correctly.

2.8 The function of the $\overline{\text{BUSREQ}}$ and $\overline{\text{BUSACK}}$ signals is to allow a device other than the CPU to control the system buses. The signals are used generally when a direct memory access controller chip is employed in the system. The DMA controller chip sends a bus request signal to the processor to indicate that it wishes to take over the system buses. When this is received by the CPU it responds with a bus acknowledge signal.

2.9 The microprocessor is the only 'talker' on the address bus, indicates that this is the only device that can produce information for the address bus which is then used by other devices in the system. This indicates that the address bus is uni-directional.

2.10 Tri-state devices are required for connection to the data bus to eliminate the possibility that two conflicting bits of data will be presented simultaneously on the data bus by different devices. Only one device is permitted to present data at any moment and all of the devices must either be switched into their high-impedance state, or their read state.

2.11 This system has no bus drivers, the number of chips connected to its address bus must be relatively small, within the fan-out capability of the microprocessor.

CHAPTER 3

3.1 The 68000 uses TTL level signals and a 5 V power supply, and therefore can be considered to be fully TTL compatible.

3.2 The 68000 microprocessor has no I/O request line because it only employs memory mapped input/output.

3.3 The Z80 has 16 address lines and can therefore directly address 64 Kbytes of memory.

The 68000 microprocessor has 23 address pins, which make it possible to address 8388608 words of memory. However, each word is 16 bits, which give it an effective addressing range of 16777216 bytes (16 megabytes). The 68000 has no A_0 since this is not required by the memory but it is used internally in the microprocessor.

3.4 The maximum power dissipation of the 68000 is given as 1.5 W. Therefore since it has a single 5 V supply, the maximum current requirement would be 300 mA.

3.5 The main advantages of using a dedicated serial input/output chip for serial data transfers are that all the logic that needs to be implemented comes in one chip. It also includes automatic error correction and checking so that additional circuitry is minimal. Programmable SIO devices can generally be provided that allow data transmission in a number of formats which would otherwise require a complete redesign of a digital logic circuit.

The only real disadvantage is that if very simple serial input/output is required, they would provide an expensive solution, and may be over-complicated to program.

3.6 The CLK input is the system clock that would be connected to the microprocessor clock so that the device can operate synchronously with the data bus in the system. However, data transmission takes place at a completely different rate and this is determined by the transmitter clock TxCA. This clock may vary in frequency depending upon the rate of data transmission selected.

3.7 INIT: LD A,0FFH
             OUT (82H),A   ;   PORT A MODE 3
             out (82H),A   ;   ALL BITS INPUT
             OUT (83H),A   ;   PORT B MODE 3
             LD ,0FH
             OUT (83H),A   ;   BITS 0–3 IN, 4–7 OUT
             . . .

3.8 The mask control register contains only 2 bits.

One determines whether an interrupt will be generated on the high or low level of the data, and the other determines whether **AND** or **OR** logic will be used for multiple bits which could cause an interrupt.

CHAPTER 4

4.1 400 mW.

4.2 7.0 V.

4.3 t_{DSW}, which for the MK4118–1 is 20 ns.

4.4 All of the MK4116 logic signals are TTL compatible. However, its power supplies are not since it requires +12 V and −5 V in addition to the normal +5 V.

4.5 The 2716 has the capability for single location programming, but this only applies where a logic '1' in a bit must change to a logic '0'. Since the data was written as 55 hex, i.e. 01010101, and it should have been 54 hex, i.e. 01010100, it should be possible to change the logic '1' in bit 0 to a logic '0' using the single location programming facility. This would require a programmer that was capable of delivering 150 ms pulse to the required address with the correct data applied to the data bus.

4.6 The program operation table gives the programming supply current as 30 mA. It would therefore be sensible to design a power supply to deliver at least 50 mA.

4.7 The MK36000 automatically switches to stand-by mode when the CE pin is held at a TTL HIGH level.

4.8 The MK4116 dynamic memory is completed refreshed every 128 memory cycles. The actual time this takes will depend upon how frequently the memory cycles occur, which in turn depends upon the processor clock speed. However, the complete device must be refreshed within 2 ms otherwise it will lose its data.

4.9 The OE pin on the MK4118 activates the I/O buffer into its output mode, whenever the device is required to provide data for the system. The chip select CS pin must be active whenever the device is required to either **read** or **write** data.

CHAPTER 5

5.1 Polling would be preferred in a microcomputer system if it was difficult to include additional interrupt logic. It may also be preferred if the data input from a number of peripherals occurred on a regular basis and the time occupied by the polling algorithm was negligible.

5.2 An interrupt facility in a microprocessor system

allows the CPU to respond very quickly to requests for input or output. Interrupts eliminate the need for regular checking of peripherals by the CPU.

5.3 The contents of the registers must be saved during an interrupt service routine so that the processor can return to the main program and continue with its operation as though it had not been interrupted.

5.4 When an interrupt occurs further interrupts are **disabled**. Therefore an enable interrupt instruction is required during this service routine to allow future interrupts to occur. This is the only method by which future interrupts can be re-enabled.

5.5 The main difference between the maskable and a non-maskable interrupt is that a maskable interrupt can be prevented by the software, whereas a non-maskable interrupt cannot. Generally the two types of interrupt have separate pins on the CPU to accept the signal.

5.6 The only maskable non-vectored interrupt occurs if the Z80 Mode 1 is used. Therefore the required instructions would be:

IM 1

EI

5.7 A vectored interrupt allows the address of a service routine to be determined by the data which is read in on the port immediately after the interrupt occurs. A non-vectored interrupt always branches to a fixed address.

5.8 A Z80 has 8 restart instructions. They are as follows:

5.9 The instructions to initialise the CPU for mode 2 interrupts would be:

Restart	Hex code	Address
RST 0	C7	0000H
RST 8	CF	0008H
RST 16	D7	0010H
RST 24	DF	0018H
RST 32	E7	0020H
RST 40	EF	0028H
RST 48	F7	0030H
RST 56	FF	0038H

IM2

LD A,n – where small n is the **high** byte of the interrupt service routine start address table

LD I,A

EI

In addition the interrupt service routine start address would have to be stored in memory in the interrupt start address table.

5.10 (i)   the I register must contain 18.
(ii) the two ports must have interrupt vectors of 20 and 22 respectively.
(iii) The start addresses are 191B hex and 1A20 hex.

5.11 The main advantage of a daisy chain priority system is that it can be achieved without the addition of extra logic devices in a Z80-based system. Any number of devices can also be daisy chained, and no additional software is required.

5.12 Mode 0: These are 8080-type interrupts. The interrupting port must supply a **restart** number when the interrupt is acknowledged which forces the processor to **call** a specific location.

Mode 1: This is a non-vectored type of interrupt. The processor performs a **call** to address 0038 hex.

Mode 2: This is a vectored interrupt. The interrupt vector supplied by the interrupting port is combined with the contents of the CPU I register to form an address. This address and the next are examined to find the service routine start address.

5.13 MODE1: IM 1
EI

The service routine would be located at address 0038H.

5.14 (i) A non-maskable interrupt cannot be prevented by the system software. When this facility is not required the NMI pin must be held **high**.
(ii) A maskable vectored interrupt can be prevented by the system software and so an EI instruction must be executed before it can be used. The interrupting port must supply a vector which will allow the CPU to locate the correct interrupt service routine.

5.15 The interrupt register I must be loaded with 19H and the interrupt vector must be changed to 10H.

5.16 The service routine will only execute once, and then all future interrupts remain disabled.

5.17 The interrupt control byte must be changed to allow **AND** logic, i.e. change it to D7. Then the interrupt mask must be changed so that only bits 0 and 1 are monitored, i.e. change it to FCH.

The program becomes:

Address	Hex code	Mnemonic
180A	3E D7	LD A,0D7H
180C	D3 82	OUT (82H),A
180E	3E FC	LD A,0FCH
1910	D3 82	OUT (82H),A

5.18 (a) To clock the counter.
(b) To start the timer when it is programmed for an external start pulse.

5.19 36.7 ms.

5.20 Lack of pins in the 40 pin package.

5.21 The prescaler could be either 16 or 256.
If the prescaler is 16, the time constant is 125.
If the prescaler is 256, the time constant is 8.
Note that the second option provides a pulse every 1.024 ms, whereas the first provides a pulse at exactly 1 ms intervals.

5.22 LD A,00011111B
OUT (41H),A
LD A,125
OUT (41H),A
$$1 \times 10^{-3} = 500 \times 10^{-9} \times 16 \times TC$$
TC = 125

5.23 Channel 0 has the highest priority and channel 3 has the lowest. These are assigned automatically within the CTC logic.

5.24 549 755.81 seconds or 6.36 days approximately.

5.25 8 μs.

CHAPTER 6

6.1 BCD code is used in computer systems because:
(a) It can be manipulated with almost as much ease as binary.
(b) It is a convenient method to represent numbers which will be used for digital display systems.

6.2 A **software buffer** is an area of memory specially reserved to store numbers or data used during a calculation.

6.3 Each byte of binary data is divided into two 4-bit numbers and these 4-bit numbers are encoded as ASCII characters. Each pair of ASCII characters is then transmitted from one computer to another, and on reception decoded and re-assembled into the binary data.

6.4 If the analogue voltage is higher than the maximum that can be produced by the computer, and the ramp method of conversion is being used, the computer will never finish cycling the ramp since the comparator will never change state. Therefore, if a display system is being used, this will not show any reading.

With the successive approximation method, the comparator will indicate that each value is higher than the computer-generated voltage but the result will still terminate after 8 samples, with the value FF as the analogue equivalent.

6.5 The main disadvantage of the **ramp** method is that the conversion time varies according to the analogue value. The higher the unknown voltage, the longer the conversion time. In addition, the conversion time is relatively long compared with other methods.

6.6 7-segment code depends upon the system hardware connections and generally bears no mathematical relationship with other forms of data.

CHAPTER 7

7.1 A designer would use 1:1 artwork if a small scale production run was required, in which the precision was not vital. It would also provide the cheapest method of producing a single-sided board, since no photographic process is required.

7.2 Double-sided plated-through hole boards are the norm in the digital electronics industry because of the need to pack components very densely, and to minimise the number of additional wires or links that must be provided on boards. They are also easily designed by computer-aided design packages.

7.3 A computer-aided design package for PCB artwork has the following advantages and disadvantages.
Advantages
- It provides a much quicker process for large printed circuits, particularly those that are highly complex.
- Most packages demand precision in the circuit diagram and the layout diagram. This eliminates possible errors.
- Automatic track routing speeds the design process, while also allowing changes to be done very easily.
- All the required artwork can be produced by the computer, including in most cases component lists and drilling details.
- Some CAD packages can be linked directly to the automatic drilling machines used in high-volume production.

Disadvantages
- The equipment and software required are relatively expensive, particularly if a professional package is required.
- Quality of the output depends upon the precision of the plotter used.
- Most packages do not automatically route 100% of the tracks, and this leaves the designer to complete the board manually.
- The libraries supplied with many packages are not adequate to cover the full range of components available.

7.4 Electroless copper plating involves the chemical deposition of copper on the surface of a PCB which has been suitably prepared. It produces a very thin layer of copper on the surface which is sufficient to provide continuity but not sufficiently strong to undergo further processing.

Electrolytic copper plating strengthens a very thin copper layer on the board surface with a thicker copper layer by electrolysis. The electrolytic plating cannot be achieved directly onto a laminate because it is a non-conductor.

7.5 120 mΩ.

If the width is increased to 1 mm, the resistance will reduce to 60 mΩ.

7.6 6.645 pF.

7.7 'Ringing' occurs because of a mismatch between the impedances of the conductors on the PCB which form a transmission line, and the gate input and output impedances.

7.8 When two boards are separated by as much as 50 cm, the most obvious precaution to take is to include a line driver on the transmitting board and the line receiver on the receiving board. This will ensure that the transmission line effects are eliminated by cutting down any reflections due to impedance mismatches. In addition, it is worthwhile adding a ground in the form of a screened cable or a twisted ground connector with a signal lead. This will help to eliminate any potential cross-talk problems.

7.9 Voltage spikes on the supply lines may be due to inadequate decoupling of the power rail. This can be improved by the addition of ceramic decoupling capacitors at regular intervals. In addition, the ground line may be of inadequate width, in which case a redesign of the track layout may be necessary. In critical cases it may be necessary to separate parts of the circuit so that they have their own supply and ground lines which are taken to a common grounding point. This ensures that current spikes and associated voltage changes are confined to specific circuit areas.

Index

1 2 4 8 16 32 64 128 256 512 1024 2048
0 0 0 0 1 1 0 0 1 0 0 1

2352
2048
304
256
48